Acta Universitatis Stockholmiensis

STOCKHOLM STUDIES IN ENGLISH

LXI

The English Plant Names in The Grete Herball (1526)

A Contribution to the Historical Study of English Plant-Name Usage

by

MATS RYDÉN

ALMQVIST & WIKSELL INTERNATIONAL

STOCKHOLM, SWEDEN

The printing of this book was subsidized by
Humanistisk-samhällsvetenskapliga forskningsrådet

Abstract

Rydén, Mats, 1984, *The English Plant Names in The Grete Herball (1526). A Contribution to the Historical Study of English Plant-Name Usage.* Acta Universitatis Stockholmiensis. Stockholm Studies in English LXI. 110 pp. Stockholm. ISBN 91-22-00710-5.

The aim of this study, which is the first outcome of the project "The English Plant Names in Early Modern English Herbals and Floras", is to analyze the English plant names in *The Grete Herball* of 1526. *The Grete Herball*, a work of 170 folio pages, is the second earliest printed English herbal. It is a most important document in the history of English plant-name usage, but has not been submitted to a systematic analysis before. The book contains some 500 English plant names, some 25 % of which are first or only recorded there. A great many names antedate the earliest entries in the *Oxford English Dictionary*. In the Introduction the *Herball* is viewed in its botanico-historical setting and the organization and content of the book are described. In the main part of the study the frequency, provenance and typology of the English plant names in the *Herball* are analyzed. Further, there are chapters on the names as attested in the *Oxford English Dictionary* and on the continuation of the names in present-day usage. A special chapter is devoted to problems of identification and paradigmatic equivalence ("synonymy") in the analysis of plant names in early texts. Three tables are appended, one of which gives a complete list of the English names in the *Herball*, with extensive commentaries.

Mats Rydén, Department of English, Stockholm University, S-106 91 Stockholm, Sweden.

ISBN 91-22-00710-5

ISSN 0346-6272

Printed in Sweden by
Almqvist & Wiksell, Uppsala 1984

Contents

Preface

English plant names, not least their paradigmatic history, merit far more attention than has been hitherto bestowed upon them. The herbals and floras published in the 16th and 17th centuries are excellent sources for plant-name studies. They are, in fact, crucial documents for the historical study of English plant-name usage—for our understanding and assessment of continuity and discontinuity in plant-name patterning. But unlike for instance the Old and Middle English botanical texts (where scholars like Peter Bierbaumer and Gösta Brodin deserve special mention) little systematic research has been carried out here. The present analysis of the English plant names in *The Grete Herball* (1526) is the first step, as part of a project on plant names in early modern English herbals and floras, towards remedying the situation.

On the completion of this book I would like, in the first place, to express my gratitude to the historian of British field botany, Mr David E. Allen, Winchester, Professor Sigurd Fries, Umeå, and Docent Gillis Kristensson, Stockholm, who have all read the entire manuscript and suggested many valuable improvements. I also owe a debt of gratitude to Professor Sherman M. Kuhn, formerly editor-in-chief of the *Middle English Dictionary*, for kindly answering questions on points of Middle English vocabulary and to Mlle Brigitte Moreau (at the Bibliothèque Nationale) and Mr Paul Hulton (formerly at the British Museum) for help with editions of *Le grant herbier*. With the late Miss Blanche Henrey, that unrivalled authority on early British botanical and horticultural literature, I had many inspiring discussions. Docent Sverker Brorström, Stockholm, helped me, carefully and patiently, with the proof-reading. My thanks are further due to Professors Magnus Ljung and Lennart Björk for including my book in the series *Stockholm Studies in English*.

The British Library and the University Library of Uppsala have, over the years, provided me not only with the 'right' books but also with the 'right' atmosphere so essential for the study of early texts.

My stays in London were made possible by travel grants from the Universities of Stockholm and Uppsala. For the printing I have received a generous grant from *Humanistisk-samhällsvetenskapliga forskningsrådet*.

Uppsala, May 1984   *Mats Rydén*

Introduction

1. *The plant names in the early modern English herbals and floras and the historical study of English plant-name usage*

The history of English plant-name usage—popular, learned and normative—remains to be written. We tend to think of plant names, as of other words, in terms of lexical, individual items,[1] but like other linguistic elements plant names are part of systems or paradigms, i.e. sets of paradigmatically related units. One of the central tasks of plant-name research is to analyze the history of plant-name paradigms, i.e. plant-name "synonymy" or equivalence through time (cf. pp. 45 ff.). Such analyses of change and stability in plant-name usage will include, among other things, the study of the principles underlying plant-name formation and selection, incl. the rivalry between popular and learned names especially as manifested in Floras and similar literature.

To understand the structure of the contemporary English plant nomenclature and to assess it for normative purposes we must have the historical background. A knowledge of the history of plant-name usage is part of the equipment of the complete plant-name scholar. The evaluation of diachronic processes ('developments') should preferably be based on analyses of synchronic patterns.[2] As for English plant-name usage, the various stages in the history of the language have been very unevenly covered. For the Old English period we now have Peter Bierbaumer's comprehensive and detailed work (1975–79), with evaluations of previous studies.[3] Information on Middle English plant names is chiefly to be drawn from the MED and from editions of herbals or plant lists, such as those by Mowat (1887), Frisk (1949), Brodin (1950) and Stracke (1974). A pioneer work here was Earle 1880. The individu-

---

[1] In terms of etymology, regional distribution, etc.

[2] Strictly speaking, linguistic patterns can be designated as 'synchronic' only from the point of view of function and use (as situationally conditioned). With regard to their composition, the patterns or paradigms are diachronically layered. A corollary of this is that the 'explanation' of synchronic usage, i.e. of use and paradigm potentiality at a given time, is a matter of both synchronic and historical aspects.

[3] Bierbaumer seems however to be largely ignorant of the recent Scandinavian work in the field, e.g. of the studies by Rolf Nordhagen on English plant names.

al names as registered in the MED include numerous antedatings of earliest OED records (cf. p. 54). Principles of plant-name giving in present-day English are briefly discussed in works like Rayner 1927, Fisher 1932–34 and Dony *et al*. 1974. Grigson 1955 (1975) contains extensive lists of local names. The standard reference work on English plant names is still Britten & Holland's dictionary of 1878–86, which had been foreshadowed by that of Prior in 1863, the first book entirely devoted to English plant names. It contains however comparatively few local names.[4] Both Prior and Britten & Holland include introductory sections on the provenance and formation of English plant names. Prior's dictionary is also etymological, whereas that of Britten & Holland is only partially (and inconsistently) so. Grigson's selective etymological dictionary from 1974 (with information on the introduction into England of the species concerned and with dates of first record of the plant names treated) is a welcome addition to the meagre literature on English plant names, though not reliable on all counts.

Whereas we have a not inconsiderable knowledge of plant-name usage in the early and late periods of the history of English, remarkably little research has been done on the intervening time span (c. 1500–1900), a distribution of knowledge paralleled in other areas of English linguistic usage. This implies, for example, that none of the great herbals or the floras of the early modern period (c. 1500–1700) have been systematically analyzed with regard to the plant names they contain, albeit from other aspects, chiefly botanico-historical and botanico-medical. However, the herbals and early floras are not only corner-stones in the foundations of botanical history. They are, in addition, important, *linguistic* documents. No period in the history of English is as rich in literature mirroring popular *and* learned plant-name usage as are the three centuries or so leading up to the modern taxonomic handbooks, for instance the well-known ones by Bentham (1858) and Bentham & Hooker (1887). But apart from being consulted by compilers of dictionaries (OED, Britten & Holland, Grigson and others), this literature has been largely ignored by philologists and linguists (cf. Rydén 1981). Some preparatory work has however been carried out. The first printed English herbal, Banckes's herbal of 1525,[5] was reprinted, with plant lists and a commentary, in 1941, as were, in 1965, the early botanical works (1538, 1548) by William Turner—with introductory matter by W. T. Stearn and others, and with lists of "synony-

---

[4] Cf. Prior 1863 (p. xxi): "Provincial words that have not found their way into botanical works are, with a very few exceptions, omitted".

[5] The title derives from the name of the printer. Banckes's herbal is the first book printed in England which is devoted solely to plants. John Trevisa's translation of Bartholomaeus Anglicus, *De proprietatibus rerum* (printed c. 1498) includes one chapter on plants and trees. This work is now available in a modern ed. (1975). Cf. Henrey 1975 (1), p. 13.

mous" plant names.[6] On the English plant names in Gerard's herbal (1597) there is an introductory essay by Rydén (1978 b).

The botanical works by Turner (1538, 1548, 1551–68), Lyte (1578), Gerard (1597, 1633), Parkinson (1629, 1640) and Ray (1670, 1690) constitute the most important sources of English plant names in early modern times. For the 18th century, Hudson's flora (1762) is a very valuable document, as are the first floras in English (see p. 24).

As hinted above, the herbals are storehouses of old, popular plant names. But they also include a great many names created by the herbalists, who were linguistically conscious and who took an active interest in their native tongue. New species were discovered,[7] bringing with them an increased need for English names, names that were often modelled on the pattern of foreign names (see Rydén 1978 b). It should be noted that the herbals—at least some of them like those by Lyte and Gerard—were among the most widely read books of the day and influential on several levels of plant-name usage.[8] The early floras by Ray, Hudson and others representing the initial stages of a normalized nomenclature are particularly interesting for the light they shed on the (consistent) introduction of native "systematic" names,[9] i.e. names closely related to the scientific nomenclature, like *meadow fescue* and *sheep's fescue* for *Festuca pratensis* and *F. ovina*, respectively. Most of these names, in particular those for grasses, sedges and rushes were devised to fill gaps in the popular nomenclature. The use of "systematic" native names in the floras is however foreshadowed in the herbals, for instance in Gerard's *medow grasse* ("Gramen pratense"), *small medow grasse* ("Gramen pratense minus"), *great Foxe-taile grasse* ("Gramen Alopecuroides maius") and *small Foxe-taile grasse* ("Gramen Alopecuroides minus"). On the employment of Latin binary names and polynomials in early botanical literature, see Green 1927 and Stannard 1974.[10]

The Renaissance herbals and the early floras are crucial links in the develop-

[6] Earlier reprints date from 1877 (the *Libellus*) and 1881 (*The Names of Herbes*). William Turner (c. 1510–68) was, as far as we know, the first to study the English flora and to collect (and devise) English names for plants in any systematic way. He was "the true pioneer of natural history in England" (Raven 1947, p. 127).
[7] In e.g. Turner's works there are 238 new species for England; in Gerard's herbal there are 182. See Rydén 1978 b, p. 143.
[8] Cf. Debus 1978 (p. 49): "There is little doubt that herbals were among the most popular books printed in the sixteenth and seventeenth centuries".
[9] For the term *systematic name* as used here, see Fries 1975. On the early development of scientific binomial nomenclature, see Heller 1964. There were no generally accepted scientific plant names until the publication of Linnaeus' *Species plantarum* in 1753. Linnaeus cites as a rule Bauhin's *Pinax* (1623) as his earliest authority.
[10] Cf. also e.g. the "Index Latinus" in Gerard 1597 and below sub "Arrangement of the plant names".

ment of English plant-name usage. They are links between the (medieval) past and the present, mirroring traditional as well as new principles in plant-name formation and application.

2. *The significance of The Grete Herball*

In 1981 a project was launched at the English Department, Stockholm University, with the aim of investigating English plant names as found in the early printed English herbals and in the first English floras.[1] The present study of the English plant names in *The Grete Herball* of 1526 (GH) is the first outcome of this project. The work selected is the second earliest printed English herbal. It has not been edited or reprinted in modern times (for the 1941 ed. of Banckes's herbal of 1525, see above). It is of course not necessary to start "from the beginning" in a project like this, but the chronological procedure has the advantage of securing the relevant linguistic background for each new object of investigation and of showing how far each writer has been indebted to his predecessors. And it enhances our possibilities to assess the relative chronology of plant names and the 'development' of plant-name usage during the time under consideration.

The Grete Herball is a comprehensive work of 170 folio pages. It contains more than 500 English words for plants, fruits or spices (for the term *English* as used here, see p. 31). The importance of the work as a source of plant names was emphasized by Agnes Arber more than 70 years ago: "This book throws an interesting light on the early names of British plants" (Arber 1912, p. 45). In fact, the greatest scholarly value of the *Herball* lies in the names of plants it supplies. Maybe we are even entitled to say that "its only interest for the modern student is furnished by the listing of early English names of many plants" (Hoeniger & Hoeniger 1969a, p. 17).[2] Many of these names are of course recorded in the OED, Britten & Holland and other dictionaries, but the present study is the first attempt at analyzing the English plant names in *The Grete Herball* systematically. A considerable number of English plant names have their first recorded occurrence in the *Herball*, in some cases antedating the previously known earliest attestation by 100 years or more (cf. pp. 35 ff. and 54 ff.).

The Grete Herball is basically a medieval document (see below sub 4) and it largely mirrors English plant-name usage, popular and learned, as prevailing at the end of the Middle English or beginning of the modern period—before

[1] Cf. Fries 1977 (p. 3) and 1980 (pp. 28 f.) for a similar project on Swedish plant-name usage.
[2] Cf. also Hulton & Smith 1979, p. 19.

the great herbalists and botanists (starting with William Turner) began to set their marks on the English plant nomenclature.

The English plant names in the *Herball* represent no doubt in a good many cases genuinely current usage.[3] However, since the book is a translation of a French work (though with numerous additions and deviations),[4] many "English" names in the *Herball* are only translations or adaptations/adoptions of names occurring in the French work. The translator apparently had the ambition to give one or more English equivalents for most plant names found in the original.[5] The French text served, as it were, as a trigger of the potentialities of English plant-name formation at the time. In other words, the fact that *The Grete Herball* is (largely) a translation enhances in some respects its value as a plant-name document.

In sum, the onomastic evidence of *The Grete Herball* is significant on basically three counts: (1) as a reflex of late medieval or early 16th century plant-name usage, (2) as an indicator of the potential for English plant-name formation around 1500, and (3) as a document accessible to and utilized by the later name-recording and name-giving herbalists (cf. p. 56).

The purpose of this study is to give a contribution to the history of English plant-name usage (including plant-name chronology) by accounting for the frequency, provenance, typology and "synonymy" (i.e. referential identity) of the English plant names in *The Grete Herball*. Special chapters are devoted to additions and antedatings to the OED and to the continuation of GH plant names in present-day usage. Hence the analysis is carried out both at a synchronic and a diachronic level. The section on "the organization and content of the *Herball*" is not meant as a full-scale description. This is hardly necessary here since the book is written in a well-known herbalistic tradition.[6]

3. *Provenance and history of the Herball*

The sources

The immediate source of *The Grete Herball* ("The grete herball") is a French work commonly known under the title of *Le grant herbier en francois* though

---

[3] On the PE currency of the English plant names as given in GH, see "Plant Names" 6.
[4] See below sub "Provenance and history of the *Herball*".
[5] Cf. here Wessén 1924, p. 62. In many cases the English name represents a textual transference, as *sanguinary* for *Capsella bursa-pastoris*; *sanguinary* is on pre-GH record only for *Achillea millefolium*. But sometimes the English equivalent given in GH—as *shepeherds purs* for *bource a pasteur* (i.e. *Capsella*) in the French original—is found in sources prior to GH and may have been known to the translator independently of the French text. It is essential to remember, in this context, that the motives of popular plant names tend to be international.
[6] See e.g. Anderson 1977, Arber 1938 and 1953 and Hoeniger & Hoeniger 1969 a.

when first published (c. 1486–88) it was named *Arbolayre*. Two other incunabula editions of the French work are known to exist, dating from c. 1498 and c. 1498–1500, respectively, and more than twenty 16th century editions are attested (see Anderson 1977).

Le grant herbier ("Translate de latin en frācois") is "an expanded version of the 12th-century manuscript *Circa instans*" (Anderson 1977, p. 101).[1] It is, like most medieval herbals, an anonymous work. Most of the editions of *Le grant herbier* have around 475 chapters, of which 264 were taken from the *Circa instans*,[2] but some of the later 16th century versions are 500 or more chapters in length.

As far as we know, *The Grete Herball* is the only translation of *Le grant herbier*. Although fundamentally a translation, the English work is not merely so—it "differs frequently from its model" (Anderson 1977, p. 99)—though we do not know what edition(s) of *Le grant herbier* the translator used.[3] The name of the translator is similarly unknown. The edition of *Le grant herbier* basically used for the present study is a copy in the British Library, London, dated in the British Museum General Catalogue of Printed Books as "1500?", which is probably a copy of the third edition (cf. above).[4]

The first edition of *The Grete Herball* (1526) contains 505 chapters, whereas the edition of the French work used for the present study has 476 chapters. The most substantial GH-addition is the section called "Here after foloweth a rehersall of dyuers chapters . . ." covering 24 chapters. The preface (prologue) to the *Herball* bears resemblance to that of *Der Gart der Gesundheit* (1485), sometimes called the "German Herbarius", whereas that of *Le grant herbier* resembles that of its main source, the *Circa instans* (Anderson 1977).

For differences in the illustrative material and in the organization of plant names between the French and the English works, see pp. 16 and 26, respectively.

---

[1] *Circa instans* was written at Salerno, Italy, around 1140. It is "essentially a collective work" (Anderson 1977, p. 46). Earliest extant MSS date from about 1200. It was first printed at Venice in 1497.
[2] The provenance of *Le grant herbier* is a complicated matter, presenting an intricate complex of bibliographic problems. For relevant literature, see Anderson 1977.
[3] Of the numerous (but very scarce) 16th century eds. of *Le grant herbier* only one pre-1526 ed. (?1513) and two post-1526 eds. (?1540 and 1545) have been available to me. For significant differences as regards plant names between these eds. and the incunabula eds., see p. 38. Certain plant names in GH (unrecorded in OED and MED), such as *gangelon* and *remcope*, are not in the eds. of *Le grant herbier* available to me and might have been adopted from other eds.
[4] The British Library has recently acquired a copy of the original ed. (the *Arbolayre*), dated "ca 1486", the plant nomenclature of which seems, in all essentials, to agree with that of the "1500" ed.

14

The editions

The Grete Herball was first "imprentyd at London in Southwarke by me Peter Treueris/dwellinge in the sygne of the wodows. In the yere of our lorde god .M.D. xxvi. the xxvii. day of July." (Colophon).

Three editions followed (1529, 1539 and 1561).[1] The 1529 edition, also printed by Peter Treveris,[2] is essentially a reprint of the original edition, whereas the two other editions, printed in London by Thomas Gibson and John King, respectively, differ in various details both from the first edition and between themselves.

The edition of 1539 ("newly corrected") contains "A table after the latyn names of all herbes" and one "after the Englysshe names of all herbes", replacing "The registre of the chaptrees" of the 1526 edition. It also includes one page (placed before the table of the Latin names) where the printer directs himself to the reader ("The prenter to the reder"). The 1539 edition has no illustrations. The 1561 edition ("newlye corrected and diligently ouersene") contains only two plant illustrations (of the mandrake). The most interesting feature of this edition is that it supplies some English plant names which are not in the other editions (see pp. 29 and 35).

The illustrations

The Grete Herball is the first illustrated book on plants printed in England. Some earlier MS herbals do, however, contain illustrations—for example the Old English translation of the Latin *Herbarium Apuleii Platonici* (see Frisk 1949, p. 12, and Rohde 1922, p. 9). The humanist tradition counteracted the use of pictures (cf. Debus 1978, p. 43). The mid-16th century saw, however, a renewed union of botanist and artist, the new era in herbal illustration starting with the work of Otto Brunfels (1530–36). His artist, Hans Weiditz, was a pupil of Albrecht Dürer's.[1] The pictures in manuscripts and in herbals published before 1530 were, with few exceptions, "stylized, decorative, and

---

[1] See Henrey 1975 (1), p. 249. For three ghost eds. (1516, 1525 and 1527), see Rohde 1922, p. 207, and cf. the years of publication of GH as given in OED. In Britten & Holland 1878–86 GH is often referred to as "Trev(eris)". Banckes's herbal (1525) was reprinted, under various titles, at least a dozen times between 1525 and 1560 (see Brodin 1950, pp. 26 ff., and Larkey & Pyles 1941, pp. x ff.). It was a handier (72 quarto pages) and a less costly (unillustrated) book than the magnificent GH.

[2] For other works issued by Peter Treveris, see DNB, Vol. 57 (1899), p. 212.

[1] Weiditz' original work was in watercolour (Lawrence 1965, p. 14). As stated by Arber (Arber 1953, p. 323 (cf. also here Green 1927, p. 404), "Brunfels' work is the first in which it is possible to identify [from the pictures supplied] a high proportion of the species enumerated". On the history of botanical illustration, see Blunt 1950, Blunt & Raphael 1979, Hulton & Smith 1979 and Nissen 1966.

often recognizable only if one already knew by sight the plants they were supposed to represent" (Reeds 1976, p. 529).

The 1526 edition of the *Herball* has 477 woodcuts, including some 70 not representing plants or fruits. *Le grant herbier* comprises only some 300 pictures, many of which are less distinct than those in the *Arbolayre*. The illustrations in the *Herball* are generally still cruder than those in the French work. They are inferior—debased and reduced—copies (in black-line frames) of the figures in the "German Herbarius" and other incunabula.[2] On the continuous degeneration in plant illustration in the late Middle Ages, cf. e.g. Fischer 1929, pp. 108f., and Morton 1981, p. 154 (note 28).

In the English book there are several discrepancies between plant and picture, occasionally notified by the printer, as at "De Boragine" (in the margin): "Nota ye pictour of bōbax & borago ye one is put for ye other." In many cases, as with rice and wheat, one figure serves for different plants. On the whole, the pictures are debased (by bad copying) beyond recognition and are thus a poor instrument of verification, i.e. of little avail for the identification of the plants described in the text.[3] Plants reasonably faithful to nature in the *Herball* are, for example, the clover, the rose, the strawberry and the water-lily.

4. *The organization and content of the Herball*

Organization

The general organization of *The Grete Herball* (1526) is as follows:

Title-page including a picture (woodcut) representing "a man picking grapes and another filling a basket with herbs" (Henrey 1975 (1), p. 17). In the lower corners there are two figures representing a male and a female mandrake. The text informs us of the purpose of the book: "*The grete herball whiche geueth parfyt knowlege and vnderstandyng of all maner of herbes & there gracyous vertues whiche god hath ordeyned for our prosperous welfare and helth ...*"

Preface. At the end of this preface it is stated that "this noble worke is compyled/composed and auctorysed by dyuers & many noble doctours and expert maysters in medycynes ...".

---

[2] See Henrey 1975 (1), p. 17. Also in other 16th century English herbals the pictures were chiefly taken from Continental works, e.g. from Fuchs (Turner and Lyte) and Tabernaemontanus (Gerard).
[3] Cf. e.g. Dürer's picture of the columbine (reproduced in Arber 1938), which was drawn —"with an unrivalled combination of artistic charm and scientific accuracy" (Arber 1938, p. 204)—in the same year (1526) as GH was published.

Table of contents called "The registre of the chaptrees in latyn and in Englyssge".

Picture of a skeleton of a man. This picture is not in the other eds. of GH, nor in *Le grant herbier*, which however includes a picture (placed after the title-page) representing the presentation of a book to a cardinal.

The body of the book containing 505 chapters chiefly on plants, arranged by the Latin names in roughly alphabetical order.[1] These chapters include a section of 24 items called "Here after foloweth a rehersall of dyuers chapters whiche before hath not ben specyfyed concernyng dyuers causes of medycyns nedefull to the behofe of man".

A section on urinoscopy (which is not in *Le grant herbier*) headed: "Here after foloweth the knowlege of ye dyuersytees & colours of all maner of vrynes through the whiche the Phycysyens mynystre or cause to be mynystred all maner of medycynes to the vtyll & profytable helthe of man."

On the use of urinoscopy for diagnosing and treating of diseases, cf. Brodin 1950, p. 40.

A section on difficult words called "The exposicyō of the wordes obscure and euyll knowen" (chiefly definitions of diseases).

A quick guide to remedies for various diseases headed "Here after foloweth a table very vtyll and profytable for them that desyre to fynde quyckely a remedy agaynst all maner of dyseases …" (in *Le grant herbier* the corresponding section is placed before *Le table des herbes* = "The registre" in GH).

At the end of this guide (or index) it is stated: "Thus endeth the grete herball with his tables which is translated out y^e Frensshe in to Englysshe".

The last leaf, on which are stated the place and date of printing and the name of the printer. This leaf also contains a whole-page woodcut representing "a wild man and woman holding up a shield. On the shield are the printer's initials and mark, and beneath the figures is a ribbon with his name" (Henrey 1975 (1), p. 18).

The description of a plant in the *Herball* may include the following items (cf. below sub "Plant lore and herbal medicine" and "Botanical description"): "degree" (hot, dry, etc.), "kinds", habit of growth, habitat, time of gathering and use of the plant. Sometimes information (in quite general terms) is supplied on the frequency and provenance of a plant. For the arrangement of the plant names (English and foreign), see pp. 26 ff. The discussion of the medical and culinary use of a plant invariably occurs as last item, usually

---

[1] As in other medieval books on plants, "the alphabetical order is usually not carried beyond the first letter" (Stannard 1974, p. 24). The alphabetical order also obtains in Banckes's herbal (1525) and in Turner 1538 and 1548, but not in Turner 1551–68, Lyte 1578 or Gerard 1597. But still in Ray 1670, genera are arranged alphabetically.

under a separate heading. The section on the medicinal properties of the plant is only occasionally missing, as for instance with *Ferula* and *Pes leonis*.

Plant lore and herbal medicine

Although published in 1526, *The Grete Herball* is, in organization, content and style, a medieval document. It is a mixture of facts and fancies, a reflex of both rationalism (empiricism) and irrationalism (superstition), testifying to the belief in the mystery of herbs and in their healing and destroying properties. In no respect does it foreshadow the expanding knowledge of botany in the 16th century. The concern of the 16th century English herbalists to find equivalents in the English language for foreign plant names is however also evident in the *Herball*.[1] In the Middle Ages the basic reason for studying plants was to learn their medicinal virtues and culinary uses. And, like its predecessors, the *Herball*[2] is essentially a practical guidebook designed to supply useful advice on the medicinal and pharmacological properties of plants, plant products, animals, metals, etc. Consequently, the plants mentioned in the herbals are almost exclusively those considered useful (or dangerous), which of course markedly limits our knowledge of the actual floras of the time and of the native nomenclatures. The medieval garden, too, "was essentially utilitarian in lay-out, contents, and intention" (Morton 1981, p. 151). Throughout the 16th century and even later the herbal remained popular chiefly as a work on *materia medica*.[3]

The medico-historical background of *The Grete Herball* is well known. The information provided on the medicinal use of plants is that found in any medieval book on plants (cf. e.g. Brodin 1950 and Frisk 1949) and need hardly be exemplified here. It includes, as usual, the classification of substances into different "degrees", defined in the *Herball* in the following way (sub "The exposicyō of the wordes obscure and euyll knowen"): "Degre is the quantyte in the which the pacient or seke body is hote/colde/drye or moyst/and there be .iiii. degrees in medycynes ...".[4] Judging from the space devoted to them in

[1] Cf. above, p. 13, and Stannard 1974 (p. 26): "Much emphasis was placed upon vernacular names and synonyms, for this was the natural means of communication".
[2] The term *herbal* is not found in medieval records. The earliest ex. of *herbal* in OED derives from the title of GH (see also Lawrence 1965, p. 3). The term *flora* 'descriptive catalogue of all the plants of a particular area' is first attested in the 18th century.
[3] The first floras, too, contain some herbalistic matter, like Ray 1670/1690 and Withering 1776. It was not until the 20th century that all medicinal references were eliminated from the floras.
[4] Cf. the definitions of *degree* as given in a modern work like Anderson 1977 (p. 245): "A noticeable difference in the strength and effect of a medicinal substance. Rated in terms of hot or cold, moist or dry, the degrees ranged from temperate (perfectly neutral), to mild (or first degree), to the strongest possible (or fourth degree). Items in the fourth degree were usually poisonous or caustic". For the terms *degree* and *humour* as used in the old herbals, see also Larkey & Pyles 1941, pp. xix ff.

18

the *Herball*, the betony, the plantain and the wormwood seem to have been considered the most efficacious plants, with the betony in the lead (the description of that plant includes 39 items). As pointed out by Hoeniger & Hoeniger 1969 a (p. 17), "not all the recipes in the work are strictly for illness. Some are designed to guard one against forgetfulness, others help one to be merry" (cf. below). In fact, "For forgetfulnesse" is a recurring subheading in the *Herball*.

Often a note on the right time of gathering a plant is given, as with brotherwort ("It ought to be gadred whan it bereth floures"), crowfoot ("it may be gadred at all tymes"), dropwort ("the rote . . . ought to be gadred in heruest tyme/and may be kept X. yere in strength"), herb John ("It ought to be gadred in June or July whan it floureth and hanged in a shade to drye") and (the berries of) juniper ("This sede ought to be gadered whan in heruest and may be kept two yeres"). With certain plants, some superstition was associated with the collecting (cf. Høeg 1976, pp. 110 ff.), as with case-weed ("It ought to be gadred in June/in the waynynge of the mone") and vervain ("take ye powdre of this herbe that was gadred whan the sonne was at the hyest"). About the vervain it is also stated that "to make all them in a hous to be mery take foure leues and foure rotes of vernayn, in wyne/than sprvncle the wyne all about the hous were the eatynge is and they shall be all mery".[5] More rarely, other folk customs or beliefs associated with plants are mentioned. Mullein (*Verbascum*) "is made a maner of torches whan it is greased" and sub "De Siligo" a good deal of advice is given as regards various kinds of bread, as also a *caveat*: "Take hede of eaten [i.e. eating] al maner of brede y^t is not baken wel for it causeth many dyseases in the body". "Rye", it is said, "nouryssheth more than the barly", whereas wheat "ought to be gyuen to labourers".

The discovery of mugworts was attributed to Diana: "It is to wyte that Dyana founde these thre mugwortes/and theyr vertues and she gaue this same herbe to Centaurus/whiche proued the vertues therof many tymes/and therfore Dyana named it arthemesia" (sub "De Arthemesia minima"). Compare Shakespeare's plant name *Dian's bud* (Rydén 1978 a, p. 76).

Certain plants were valued for ornament. An example is *Calendula*, the marigold, about which it is said that "Maydens make garland of it whan they go to feestes and brydeales [i.e. bridals] bycause it hath fayre yelowe floures and ruddy". Herbs could be used as amulets, for instance the vervain, about which it is stated that "the rote of this herbe hanged about his necke profyteth moche . . . For payne of the heed were [i.e. wear] a garlande therof for it

---

[5] Cf. here the use of mugwort: "To make a chylde mery/hange a bondell of mugwort . . . or make smoke therof vnder the chyldes bedde/for it taketh away anoy for them" (sub "De Arthemesia minor").

taketh ye heate away meruayslously ... Agaynst bytynge of serpentes or other venimous bestes/who so bereth this herbe in his hande or hath it gyrde about hym shall be sure of all serpentes".[6]

The old use of orchids as aphrodisiacs is mentioned sub "De satirione": "At the rote be two thynges as ballokes that be good in medycyns/whan they be grene they be confyct with hony/and aydeth lechery".[7]

One of the central plants in old herbals is the mandrake (*Mandragora officinarum*), a plant which is linked with a number of superstitions (cf. e.g. Brodin 1950, p. 236). In *The Grete Herball* we find however certain disbeliefs in the alleged features of the mandrake (cf. Arber 1938, p. 123, Brodin 1950, p. 184, and Rohde 1922, p. 73): "Some say that the male hath fygure or shape of a man. And the female of a woman/but that is fals. For nature neuer gaue forme or shape of mankynde to an herbe. But it is of troughe [i.e. truth] that some hath shaped such fygures by craft/as we haue somtyme herde say of loboures in the feldes". The passage is translated *verbatim* from the French.

It should finally be pointed out here that the plant names themselves have a good deal to tell us about popular attitudes to the plant world, the motivation behind popular plant-name formation being basically twofold: (1) the characteristics (incl. the habitats) and uses of plants and (2) fanciful ideas about plants. For the classification of plant-name motives, see, for example, Britten & Holland (pp. xvi ff.), Lange 1966 and Vide 1967.

Botanical description

A consequence of the markedly utilitarian character of the herbals is that the plants mentioned are chiefly those of use or of alleged use to man. As remarked above, this implies a constraint on our knowledge of the current names for plants at the time concerned, though few plants other than those of value to ordinary people (or troublesome plants) were given popular designations (cf. Prior 1963, p. viii, and below sub "Plant Names" 4). Few plants (if any) had names for merely aesthetic reasons (see e.g. Hesselman 1935, p. 86, Høeg 1976, p. 683, and Rydén 1981). Another limitation on our knowledge of medieval vernacular plant names is the fact that the medieval herbals reflect an international, herbalistic flora, including "exotic" plants, chiefly from the Near East, rather than country-specific floras. Little, or at least no consistent, distinction was made between exotica and indigenes.

---

[6] Cf. Rohde 1922 (p. 29): "The herbs in commonest use as amulets were betony, vervain, peony, yarrow, mugwort and waybroad (plantain)."
[7] Cf. Gerard 1597 (p. 158): "Our age vseth all the kindes of stones [i.e. orchids] to stirre vp venerie, and the apothecaries doe mixe any of them indifferently with compositions seruing for that purpose".

Problems of organization and classification were of little concern to the medieval (post-Classical) herbalist, at least as reflected in the herbals: the plants were simply arranged in (roughly) alphabetical order in the sequence of the Latin names.[1]

In their praiseworthy survey of "the development of natural history in Tudor England" (1969) Hoeniger & Hoeniger state (pp. 16 f.) that *The Grete Herball* "contains almost no botanical information whatsoever but was designed purely for physicians as a handy compendium of herbs and their traditional remedies". However, even with allowances made for possible definitions of "botanical", this statement is hardly correct. Like many earlier herbals (see e.g. Brodin 1950), the *Herball* provides information, however imprecise, on quite a few botanical aspects. Since the *Herball* does not deviate essentially here from other medieval or early 16th century herbals, only a few quotations are needed to give an idea of the kind of information supplied and the rather artless language in which it is conveyed.

(*a*) Provenance and frequency

The home of exotics is sometimes given as "Babylon"[2] or "beyond the sea" (corresponding to *oustre mer* in the French text) or in similar terms:

alkanet ("This herbe is founde in places beyonde the see/and specyally in Cyryll"); balm tree ("it is founde towarde Babylon"); Cassea ligna ("It is the barke of a lytell tree that groweth towarde the ende of babylon"); cotton ("Bombax is cotton and is an herbe that groweth beionde the see/and in Cycyll is grete quantyte"); aloes ("This wood is founde in a flode of hye Babylone nygh wherby renneth a ryuer of Paradyse terrestre").

Some other geographical indications: Agaricus ("in Lombardy"); camel's straw ("in arabye and affryke"); Culcacia ("groweth moost in Egypte"); nenuphar ("in all regyons hote and colde/but the best is in a hote regyon"); cloves ("in ynde").

If at all indicated, the frequency of a plant is usually stated in terms of "common" (e.g. *Daucus*), "very common" (e.g. *Calendula*), "common enough", "abundantly", "in great quantity", etc., usually following the French text. More occasionally, frequency is indicated in terms of rarity, as with *Daucus creticus*: "it is not moche founde here". For *common* as a plant-name epithet, see below sub "Plant differentiation".

[1] Cf. above sub "Organization" and Earle 1880, pp. xxi f. In the 16th century, there were however various attempts at the classification of plants (by Bock, Caesalpinus, Lobelius, and others.)

[2] As stated by Arber 1953 (p. 320), such references to Babylon are probably "not intended to be taken literally, but were merely a graceful way of indicating that the writer had no idea whence the plant came".

(*b*) Habitat

Since it was essential for the physician or the herbwoman to know where to find the plant in question some information about the habitats of plants is usually supplied. In fact, "the herbalists regarded an indication of the habitat as an actual part of the description" (Arber 1941, p. 21). Some quotations:

alleluia ("This herbe groweth in thre places/and specyally in hedges/woodes/ and vnder walles sydes"); case-weed ("groweth by pathes and hye wayes"); cockle ("groweth amonge wheet"); dodder ("It is an herbe y^t wyndeth about flax or lyne growynge"); duck's meat ("It is a lytell rounde wede that groweth swymmynge on ye water in pondes and styll waters"); gladdon ("it groweth not onely in water/but is also founde in hygh groundes"); houseleek ("It groweth vpon houses"); knotgrass ("It groweth in wayes and feldes"); lady's seal ("It groweth in derke shadowy places and in forestes"); oak fern ("groweth on walles/stones and vpon okes/and that on the okes is best"); oxtongue ("It groweth in very sande places"); plantain ("groweth in moyst places and playne feldes").

(*c*) Plant characteristics

In early botany, the description of a plant was based on vegetative characters, including the underground parts of the plant. The flower (apart from colour) was rather neglected and the floral structure was not considered at all; in fact the early botanists "knew nothing about the essential organs of the flower" (Green 1927, p. 404). As stated by Brodin 1950 (p. 12), "details about the leaves ... were probably the most important characteristic for the identification of the plant".

Following an old herbalistic tradition, plant characteristics are in the *Herball* often given in terms of comparison with another plant (cf. Earle 1880, pp. xvii ff., on "comparative description"):

alleluia ("hath leues lyke .iii. leued grasse and hath a soure smell as sorell"); affodylly ("It hath leues lyke leke blades"); chervil ("hath leues lyke percely"); cotula fetida ("is an herbe moche lyke to camomyll/but it hath an yll and stynkynge odour/camomyll hath a good smell"); dittany ("hath leues moche lyke to strawberyes"); hops ("rampeth in maner of an herbe called bryony"); little clote ("hath brode leues lyke nenufar"); nenuphar ("hath large leues and hath a floure in maner of a rose"); Spargula ("is lyke to warence in leues, but is lesse").

Some other descriptions:

bryony ("and it hath a grete rote"); chicory ("Whan the sonne ryseth this floure openeth/and it closeth whan the sonne gooth downe"); daisy ("The

22

colour of the floure draweth somwhat towarde reed"); dawke ("hath a large floure and in the myddle therof a lytel red pricke"); herb John ("hath many small holes in the leues and bereth a yelowe floure"); tormentil ("This herbe groweth a cubyte hygh"); tutson ("It is founde grene at all seasons"). Eryngium (sea holly), a thistle-like umbellifer, is classified (sub "De radice yringorii") as "a maner of thystle".

In many cases the description is quite fanciful due to lack of knowledge and/or confusion with other plants. An example in point is the following character-ization of the cowslip: [it] "groweth at yᵉ fete or sydes of hylles in watery places: The leues thereof be lyke leues of rew [i.e. rue] and groweth in maner of a tre".

(d) Plant differentiation

The medieval and 16th century botanists had no conception of species or variety in the modern sense of the terms. The first to differentiate between 'genus' and 'species' was Caspar Bauhin in his *Pinax* (1623). In *The Grete Herball*, plants of similar appearance are differentiated by way of general terms such as *kind* or *manner* or by contrasting epithets referring to habitat, habit of growth, season of flowering or gathering, etc. Examples of such rather loosely applied specifications (mostly calqued on the French text) are *common/wild, garden/wild, tame/wild, great/middle/small, more/mean/less, long/round, summer/winter* (cf. lists appended to the Overall list). Distinc-tions based on flower or fruit colour are also common, as with nenuphar ("It is of two maners. One is whyte/and another yelowe") and plums ("there be two sortes of them/blue and reed").

Irrespective of the specification being used attributively (as in "common garlick" or "tame onion") or predicatively, it is in most such cases difficult to decide whether or not a distinct "species" (in modern terms) is intended. And, of course, the status of the term *name* is often uncertain here.

Not a few plants are distinguished as "male" or "female" without any reference to the sexuality of plants.[3] Two examples (out of many) from the *Herball* are the mandrake (see above) and the hig(h) taper (*Verbascum*), about which it is said (sub "De tapso barbato"): "There is male and female. The femell is greter and hath broder leues/and is the better of bothe".

Of mushrooms two "manners" are distinguished (sub "De fungis"): "one maner is deedly and sleeth them that eateth of them and be called tode stoles and the other dooth not".

---

[3] See Arber 1953, p. 332. Cf. also Brodin 1950 (p. 227): "It seems that this method of attributing secondary sexual characteristics to plants was common in Greek texts and hence spread to other European scientific works". Quite often this "sexuality" in plants was associated with flower colour (cf. Lawrence 1965, p. 15).

The Herball and the English flora

Strictly speaking, *The Grete Herball* is not a book on British plants, but on plants as mentioned in *Le grant herbier*, of which some 150 may also be recognized as British, which of course does not necessarily mean "native species" as known to the translator.[1] Most probably the translator was not aware of the difference between the Mediterranean/Continental flora and that of England. Like other English herbals prior to that of Turner (1551–68), the *Herball* offers practically nothing of interest from a floristic or phytogeographical view-point.[2] There is no information about the British, as distinct from the "European", flora. Nor are there any references to British localities for plants, only some vague ones to "exotic" plants (see above sub "Provenance and frequency") and a few others, revealing the sources of *Le grant herbier*. The first work to contain references to localities for British plants is Turner 1548. The first comprehensive local British flora is Ray 1660. Ray is also the author of the first floras of the whole of England (1670/1690).[3] The native British plants as part of the *Herball* are of interest only because of the English names attributed to them there. Designations like "common" or "tame"/"wild" should of course be interpreted here essentially in terms of the French original (cf. above).

Few non-flowering plants are described in the *Herball*. Included are a few ferns (of e.g. the genera *Adiantum, Asplenium, Ceterach, Polypodium*), probably one horsetail (*Equisetum*), some "mosses"[4] and one lichen, "crayfery" or "lungwort" (the picture shows a lichen but the text describes a species of *Pulmonaria*). There is also a "sponge" and a section on corals, in those days and even later (see e.g. Gerard 1597, p. 1381) supposed to be plants. For mushrooms (edible) and toadstools (non-edible), see above. A specific kind of fungus described is agaric (*Polyporus*).

---

[1] In Turner's works some 300 native species are mentioned, in Gerard (1597) some 500. Around 1700 some 1 000 native British species were known to botanists. Today more than 2 000 species of flowering plants are found wild (or naturalized) in the British Isles.

[2] But cf. Brodin 1950, p. 25. Another matter is that some general conclusions as regards the (relative) frequency of wild plants can be drawn from the *Herball* (and similar contemporary works), e.g. that some plants like *Agrostemma githago, Cuscuta epilinum* and *Lolium temulentum* were more common in those days than today.

[3] For the earlier lists of British plants by W. How and Ch. Merret, see Gilmour-Walters 1962 (pp. 10 f.) and Smith 1824 (pp. vi f.). Ray's *Synopsis* "remained unrivalled and not superseded until William Hudson published in 1762 his *Flora Anglica* which applied to British plants the binomial nomenclature for species introduced by Linnaeus in 1753" (Ray Society ed. of the *Synopsis*, 1973, pp. 3 f.). The first floras in *English* were published in the latter half of the 18th century. See Henrey 1975 (II), pp. 119 ff.

[4] "Vsnea is of dyuers maners/some groweth on trees of good odour/as garnates/and otherlyke ... Some mosse groweth on the oken trees/and on other trees. Some groweth on stones". (sub "Vsnea. vel muscus arborum"). Another kind of moss, the liverwort, is mentioned sub "De epatica".

24

The non-botanical content

Like other early herbals, *The Grete Herball* includes a good many things besides herbs (some 100 items). Many products of plants are described, as amber, butter, cheese, cotton, hony, opium and sugar, as are various gums of trees. The utility of pitch, pearls, gold and silver (and other metals) is also mentioned as is that of glass and various "stones". Some 15 animals are referred to, such as the beaver, the fox ("a subtyll beest"), the goat, the hare, the hart, the ox, the dove, snails, spiders, worms and fishes. As for the hare, it is stated that "of all bestes is none flesshe which causeth so heuy blode and melancolye as dooth the flesshe of the hare". Spider's web is said to have the "vertue to staunche blode". "Mommye" [i.e. mummy] is described as "a maner of spyces or confeccyons that is founde in the sepulchres or tombles of deed bodyes that haue be [i.e. been] confyct with spyces". The use of water is dealt with at some length. The attitude is a rather negative one: "the maysters[1] say that water is not good to be dronken" and "many folke that hath bathed them in colde wa[ter] dyed or [i.e. before] they came home". Further, it is stated that "amonge all waters/rayne water is best" and springs "that sprynge agaynst east and south ben best/but those that sprynge agaynst the west ben the worst".

---

[1] i.e. essentially Galen and Dioscorides.

The Plant Names

1. *Arrangement of the plant names*

As stated above, the listing of the plant headings in *The Grete Herball* is roughly alphabetical according to the (chiefly) Latin name or formula, though the alphabetical order is often broken.[1] In the majority of cases the heading has the form of a *de*-phrase, for example *De Arthemesia* (in the "Registre" the prepositional formula is never used). Sometimes the Latin name cooccurs with a Greek and/or "Arabic" designation.

In some 73 % of the plant headings one or more English names are added. It should be observed that in the French work no French names are supplied in the headings (though occasionally in *La table des herbes*), only a Latin name (formula) with formal variations such as *De Agnus Castus*, *De Apio* and *Anacardus*. The plant names as given in the running text of the *Herball* are of course basically a reflex of the name-giving in *Le grant herbier*. But the *Herball* also contains a number of plant names which have no correspondences in the French original.

For a complete list of GH chapter headings containing English names (or some other botanical addition in English), see pp. 65 ff. Here a few instances may suffice to illustrate the arrangement of plant names in these headings:

De Agno casto. Tutson
De Apio risus. Crowfote or ache
De Arthemesia. Mugwort or moderwort
De bardana. A clote that bereth burres
De cicuta. Hemlocke
De Cucurbyta. A gourde
De Eruca. Skyrwyt. Or wylde cawles that bered mustarde sede
De Filipendula. Dropwort
De herba paralisi. Cowslyp or pagle
De lapaceola. Lytell burre or clyuer
De lolio. Cokyll
De Nespilis. Medlers or open arses

---

[1] In many cases an item is misplaced because of the initial word of the *text*. Two exx.: "De narsturcio" is placed after "De Scabiosa" because the text begins: "Senacions is cresses" and "De lingua passerina" is placed after "De Penthafilone" because the text begins: "Poligonia is an herbe called sparow tongue".

De sanguina[r]ia. Blodworte/or yarow
De semper viua. Howsleke or selfegrene
De sicla/alias bleta. Betes
De solatro rustice. Dwale or more morell
De viperina/alias vrtica mortua. Deed nettel or archaungell
Acorus. Gladon
Adianthos. Maydenwede
Ameos. Woodnep/or peny wort
Calendula. Mary gowles/or ruddes
Citrum. A tre so named
Squinanto. Camelles strawe
Zizania. Ray/drawke/darnell
Absinthium latine. Grece absinthion. Saxicon Arabice. Wormwood
Acetosum latine. Numa Arabice. Oxiolapatium Grece. Sorell
Allium latine. Scordon vel scordeon grece. Thaū Arabice. Gorlyke
Anisum latine & grece. Aneisum Arabice: Anys

In the text attached to the heading, the English name of the heading may just be repeated, as in:

De filice. Ferne. Filex is ferne
De Fragaria. S[t]rawberyes. Fragaria is an herbe called stra[w]bery
De fungis. Mussherons. Fungi ben mussherons
De Nigella: Cokyll. Nigella Cokyll is hote and drye in ye thyrde degre
De salicibus. A wyloue tree. Salix the wylowe is a commyn tre

In many cases, however, the appended text adds to the English name(s) of the heading. In the great majority of cases these additions are naturally reflexes of the French text—as *cuckowes brede* for *pain de cocu*, *frogges fote* for *ranoullie*, *grete plantayne* for *grant plantain*, and *sparow tongue* for *langue de passerat* (cf. pp. 37 ff.)—though independent additions occur, as *blodeworte* (sub "De Persicaria") and *moleyne* (sub "De tapso barbato"). Many of the English "parallel names" are recorded in texts prior to the *Herball*, for instance *celendyne* ("De Celydonia"), *deed nettel* ("De viperina/alias vrtica mortua"), *shepeherds purs* ("De bursa pastoris") and *sparow tongue* ("De lingua passerina"),[2] and may have been part of the translator's plant-name repertoire independently of the French text.

Some illustrative citations:

---

[2] As pointed out above, the name-giving in the text is usually determined by the French original, as in: "De lingua passerina. Sentynode. swynes grasse knotgrasse/or sparow tongue. Poligonia is an herbe called sparow tongue" (<Polligonia/cest vne herbe q̄ on appelle langue de passerat).

De Alleluya. Wood sorell or cukowes meate. Alleluya is an herbe called cuckowes brede

De bursa pastoris. Cassewed. Bursa pastoris is shepeherds purs. some call it sanguinary bycause it stauncheth bledynge of the nose

De Celydonia. Celendyne. Celidonia is a comyn herbe called Celendyne/ some call it bryght

De Diptano. Dytany. Diptanus is dytany ... Some call this herbe gardyn gynger

De Iaro. Cuckowe pyntyll. Iarus is an herbe so named ... It is also named aaron/and calues fote. Some call it prestes hode

De lentycula acque. Grenes/or ducke meate.[3] Lentylles of the water ben called frogges fote

De macianis pomis. Wood crabbes or wyldynges. Mala maciana ben wylde apples

De millefolio. Yarowe/myllefoyle. Millefoly or yarowe in some places is called carpenters grasse

De Persicaria. Arssmert or culrage. Persicaria ... is called arssmert ... Some call it sanguinary or blodeworte

De plantagine. Plantayne or weybrede. It is called also ... grete plantayne

De Saponarya. Crowsoppe. Saponaria/burit/herba fullonum herbe phylyp/ all is one. It hath many names. It is called saponary fullers grasse/buryt and crowsoppe

De tapso barbato. Hareberde/or hygtaper. Tapsus barbatus is a comyn herbe ... Some calle it moleyne/some hareberde/some hyg taper

Brancha vrsina. Bearefote. Brancha vrsina is an herbe called beares twygge or bough

Consolida media. Maythen. Consolida media is the myddle consoulde Anisum latine & grece. Aneisum Arabice: Anys. It is also called swete commyn

As appears from the above quotations, the English name(s) in the heading may be missing from the text or only one of the English names adduced in the heading may be repeated in the text. Other examples of these types of name-giving are:

De Ambrosiana. Hyndhele. Ambrosiana is an herbe lyke to eupatorium/or wylde sawge

De astula regia. Woodroue. Astula regia is a herbe so called

---

[3] In the "Registre" (sub "Lenticula aque") only *duckes meate* is adduced. The same type of reduction in variants (possibly indicating preponderance in usage) is found there e.g. sub "Millefolium", where only *yarowe* is given as an English equivalent.

Capilli veneris. Maydī here. Capilli veneris is an herbe so named

De polio montano. Wylde tyme. Polium is of dyuers kyndes

De Apio risus. Crowfote or ache. Apium risus groweth in sandy places and grauelly grounde

Zizania. Ray/drawke/darnell. Zizania is an euyll wede y^t groweth in the wheate

Calendula. Mary gowles/or ruddes. Calendula is an herbe called ruddes

De herba Indica. Gith. Cokyll. Gyth is an herbe hote and drye in ye seconde degre

In a few cases, the English name in the heading is de-modified, i.e. deprived of its epithet, or a more general term is given (usually as influenced by the French text), as in:

De Lapacio. Reed docke. Lapacium is an herbe called docke and hath many names

De bedegar. Eglentyne. Bedegart is a thorne or brere

Occasionally, the name of the heading is particularized in the text, as in:

De Altea. Malowe. Altea is a hye Malowe (<Altea est haute malue)

Some 100 or 27 % of the plant headings in the *Herball* (1526) do not include an English plant name.[4] However, in many of these cases the plant is referred to by an English (or Englished) name in the appended text. A few quotations:

De Iparis vel cauda equina. Iperium is an herbe that is called mares tayle (<queue de cheual)

De Mora bacci. Mora bacci is a wylde fruyte that groweth in busshes and breres and they be called blacke beryes

De Uulfago. Uulfago is hote and dry in the thyrde degre. Some call it hogges meate and mollum terre

De Sizania. Sizania is ray or cockyll

---

[4] Some 20 of these headings are supplied with English names (or additional English names) in the 1561 ed. of GH, e.g. De brusco (kneholme), De Cretano (waterwort), De Musis (Apples of Paradys), De Sisimbro (Baume mynte or water mynte), De Eruca (rocket added) and De Sambuco (dane wort added).

The following names in the 1561 ed. are not in the original ed. of 1526: baume mynte, bulfote, dane wort, eye bright, flewortе, garleke germander, gowlandes, horse hofe, lauender gentyll, the lesse housleeke, mastyx tree, persneps, rocket, southerwood, water germander, water mynte, waterwort.

The "Registre" of the 1526 ed. offers the following English names which are not part of the plant headings in the body of the work: agaryke (Agaricus), wylde galyngale (Ciperus), oliandre (Oliandrum), tormentyll (Tormentilla), and lychworte (sub "Gromyll milium solis").

Sometimes the French name is simply transferred to the English text (cf. pp. 31 and 40), as with *actoire*, *appolynayre* and *Croyt marine* in the following citations:[5]

Anchora. Anchora is an herbe called actoire
De appolynarya. Appolynaris is an herbe called appolynayre
De cretano. Cretanus is an herbe called Croyt marine

From the English plant names as sequenced in the headings and/or in the text of the *Herball* it is difficult to draw any conclusions as regards the frequency and currency of the names at the time. Since, however, the name-giving in the chapter headings is independent of the French original (see p. 26), we may be entitled to consider the names singled out for the headings as representing (at least in many cases) the most frequent names of the day, at any rate in the translator's opinion. The plant names appearing as part of chapter headings in the *Herball* represent some 63% of the total of the English plant names in the book.

The text accompanying a plant heading in the *Herball* often includes an array of foreign names, largely overtaken from the French text. Such "comparative synonymy" or giving of multifarious names from many languages was a way of securing the identity of a plant in plant books and of maintaining continuity with past literature.[6] This device for identifying a plant, which "has its source in Dioscorides", "continued to be the chief means to this end, down to the seventeenth century" (Earle 1880, p. xx). A few citations:

De satirione. Satirion is an herbe otherwyse called priapismus/guyos/eucar-ion/sarapias/orcis/testiculus leporis/neme/ and baram
De Arthemesia minor. The myddle mugworte is called tagantes in Grece/the domyens call it gryfauterius/y[e] Romayns tannium/y[e] Egypcyens Rym/other cal it tamaryta and other tanacipa
Aristologia longa latine. Aristologia longa is so named bycause the rote is longe/and sklender. Some cal it arratica/other melcarpon/other ephesta/other clesticis. The romayns call it petritomis
Calendula. Calendula is an herbe called ruddes. It is veray commune. It is called incuba/solsequium sponsa solis/Eulitropium/solmaria

Further illustrations of this usage in GH are offered sub, for example, "De Apio risus", "De Affodillio" and "De camomylla".

---

[5] Cf. the French text:
Anchora cest vne herbe que on appelle actoire
Appolinaris/cest vne herbe appellee appollinaire
Cretanus: Cest vne herbe qui est ainsi appellee croite marine
[6] Cf. Earle 1880, pp. cc ff., and Stannard 1974, p. 27. Although such series of synonyms may have been of some help to certain learned readers of the day, the herbalists were no doubt on the whole "much under the tyranny of names" (Arber 1941, p. 29).

2. *Frequency, provenance and typology of the English names*

A. THE DEFINITION OF *ENGLISH* AND THE OVERALL FREQUENCY OF THE NAMES

To assess the frequency of the "English" plant names in *The Grete Herball* it is necessary to define *English* as used in this study with reference to the plant names in the *Herball*. By *plant name* here is meant a word (phrase)—incl. metonyms like *blueberry* for the whole plant—which is used to identify a plant (or a fruit/spice). Like a proper name (e.g. a personal name), a plant name serves as an identification mark (or label), but like a common noun, it also subsumes particular objects under a generic concept; 'cowslip' (*Primula veris*), as a class-concept, subsumes all the features common to cowslips (cf. Ullman 1962, p. 73).

The delimitation of the concept *English* is here complicated by the non-English substratum of the English text, producing textual and nomenclatural interferences and giving us a series or scale of vernacularized (Englished) names—from translations to various types of adaptation or modification of (chiefly) French and Latin names. In addition, there are mere adoptions of (chiefly) French and Latin names.

In the present study, I have chosen to categorize the following types of GH plant names as English:

A. Names attested in English records prior to GH, except Latin designations like *esula*, *filipendula* and *palma christi*
B. (1) Names first attested in GH, as given there independently of the French text
   (2) Translations first attested in GH, as *beares bough* < *branche vrsine*, *cuckowes brede* < *pain de cocu* and *grete plantayne* < *(le) grant plantain*
   (3) Adaptations (modifications) of foreign names first attested in GH, as *coronary* < *coronaire* and *chervell* < *chèfre feuille*
C. Certain adoptions of foreign names in the French text (see below)

I do not include in the term *English plant name* direct transferences from the French text like *actoire*, *appolynayre*, *brust*, *epatyke*, *nefle*, *pederon*, *rapistre* and *rodale*, unless such names are now part of the English plant nomenclature (like *bistort*) or have been authoritatively recognized elsewhere (as in OED and/or Britten & Holland) as part of the English lexicon, as *agriot*, *artetyke*, *polytrich* and *tapsebarbe*. Nor do I include Latin names (as transferred from the French text) like *acorus*, *alchimilla*, *bursa pastoris*, *cicuta*, *esula, gracia dei, persicaria* or *tapsus barbatus*, although some of them may have been in popular use.[1]

Scrutinized according to the principles laid down above, *The Grete Herball* yields well over 500 "English" plant names (in a wide sense).[2] These names are used to refer to some 300 species or genera (cf. below sub 4).

The plant names occurring in the *Herball*, and in similar early books and plant lists, are a summation of the plant-name knowledge of the author (compiler, translator) as derived from different "sources", such as his own usage and various literary texts, native and foreign. These plant names reflect various levels of usage (difficult to delimit for lack of relevant information), along a scale from 'standard' (non-regional) names, through names of limited currency (e.g. regionally restricted names) to mere idiosyncrasies, incl. new formations and coinages.

B. NAMES ATTESTED IN OLD AND/OR MIDDLE ENGLISH

Of the some 500 English names for plants or fruits mentioned in *The Grete Herball* about 375 or 75 % are recorded in Old and/or Middle English sources. These names must, in other words, have had a certain currency, or at least some tradition, in the early 16th century, i.e. at the time of the composition of the *Herball*. The degree of currency was of course correlated with the commonness and utility of the plant(s) concerned. Further, the names were diachronically and socially layered: some were old folk names, often with a very long tradition in the language (like *cowslip*), others were just herbal or herbalists' names—in many cases translations of Latin names (like *hart's tongue*)—which were familiar to apothecaries, but presumably little known to the ordinary herbwoman.

The Old English element

Slightly over 100 of the names with pre-16th century records in *The Grete Herball* are of Old English descent, i.e. some 20 % of the total of the English

---

[1] Cf. Britten & Holland, p. 231 (sub "Gratia Dei"). For *esula* there is no English name in GH (*essell*, on ME record, appears in VBD). Cf. the lists of "English" plant names given in the 1965 ed. of Turner 1538 and 1548, where names like *filipendula* and *Cardo benedictus* are included. It must, however, be emphasized that there is no clear-cut distinction between popular and learned plant names, witness Latin names like *pyrola* and *veronica* which have become part of everyday or standard native usage (see e.g. Hesselman 1935, p. 3).

[2] Banckes's herbal (1525) contains some 250 English plant names, VBD (1527) some 260, Turner 1538 some 235 and Turner 1548 some 950 (allowing for differences in the application of *English* in the modern eds. of the works). Already in his *Libellus* of 1538, Turner records local usage of vernacular plant names.

names in the *Herball*, which testifies to a not inconsiderable knowledge of old English plant names on the part of the translator (cf. here Prior 1863, p. vi). They are in alphabetical order:[1]

alexanders, aloe, apple, ash, barley, bean, beet, blackberry, blind nettle, bonewort, box, bramble, briar, broom, brotherwort, bruisewort,[2] celidony,[3] chervil, [clarey; see below], clote, cockle, cole(wort), consoud,[4] cost, cowslip, cress, daisy, dill, dock, earthgall, elder, felwort, fennel, fern, flax, garlic, gladdon, groundsel, hemlock, hemp, hindheal, hock, hollyhock, horehound, horseheal,[5] hound's tongue, hyssop, ivy, [knee holme; see below], lavender, leek, lily, line, lion's foot, lungwort,[6] madder, maithen, mallow, mill, mint, moss, motherwort, mugwort, nettle, nightshade, oak, oats, open arse, oxtongue, pear, peony, pepper, periwinkle,[7] pine, plum, poppy, quick(s), radish, ramsons, rose, rue of the field,[8] rush, rye, savine, sengreen, sloe, smearwort, stanmarch, stichwort, strawberry, swine's grass, teasel, thistle, valerian, wallnut, wallwort, waybre(a)d, wheat, white poppy, willow, woodbine, woodruff, yarrow, yew.

GH (and ME) *clary* (Salvia sclarea) and *kneholme* (Ruscus aculeatus) are recorded in Old English as *slarie, slarige* (and similar forms) and *cnēowholen*,[9] respectively (see Bierbaumer 1975–79).

For Latinate OE forms like *fic* (Lat. *ficus*) and *lactuce* (Lat. *lactuca*), see p. 34.

Of the GH names evidenced in Old English sources slightly less than one third is of non-Germanic origin, a distribution which reflects fairly well the proportion of Germanic to non-Germanic plant names in Old English. The Old English GH names of non-Germanic origin are: alexanders, aloe, apple(?), beet, box, celidony, chervil, cockle(?), consoud, cost, fennel, gladdon, hyssop, lavender, lily, mill, mint, pear, peony, pepper, periwinkle, pine, plum, poppy, radish, rose, savine, valerian; *rue of the field* is partially non-Germanic. For the disputed origin of *apple* and *cockle*, see for instance Klein 1966–67 and Onions 1966.

Irrespective of their 'ultimate' provenance—*hyssop*, for example, is of Se-

---

[1] In this list (and the following) the names are in modern form (if any) and spelling. My chief sources for OE first records have been the OED and Bierbaumer 1975–79.
[2] OED (not in Bierbaumer).
[3] For OE forms, see Bierbaumer (especially 1975, p. 20).
[4] OED (not in Bierbaumer).
[5] See Bierbaumer 1979, pp. 140 f.
[6] For bot. ref. in OE, see Bierbaumer 1975, p. 98, and 1979, p. 164.
[7] OE *perwince* (Bierbaumer 1979, p. 184). Cf. below.
[8] OE *feldrūde* (Bierbaumer 1979, p. 89). *Rue* is ME.
[9] ME/GH and modern form by substitution of *holm* for *holen* or by phonetic change (*holen* > *holn* > *holm*). OED first record of the current form is from Turner's herbal (1562).

mitic origin—the Old English non-Germanic plant names generally came into English via Latin.[10] The acquaintance of the Anglo-Saxons with Roman plant names was a long-standing one. A few plant words, like *mint*, *pear* and *pepper*, are—as is well known—of 'Continental' heritage, belonging to the oldest layer of the Anglo-Saxon word stock. Some of the Old English plant names are translations of Latin names, a method to be developed more extensively by Medieval and 16th century herbalists and botanists. Examples in point from the *Herball* are *hound's tongue* (Lat. *lingua canis*), *lion's foot* (Lat. *pes leonis*) and *oxtongue* (Lat. *lingua bovis*).[11]

Another characteristic of Old English plant names, as reflected in the *Herball*, is that the greater number are non-compounds. For GH names of ME provenance the reverse situation obtains.

The Middle English element

More than half or about 55 % of the English plant names in *The Grete Herball* are of ME heritage, or are, more properly expressed, first attested in ME sources (c. 1150–1500).[12]

The great majority of the GH plant names of ME provenance are apparently first evidenced in the period 1200–1400. Of the GH names recorded in the printed volumes of the MED (letters A through P), some 6 % (17/18 names) are attested after 1400 (between 1400 and 1500),[13] whereas only 2 % or four names—*canel, fig,*[14] *goat's leaf, horsemint*—have their earliest attestation before 1200 (between 1150 and 1200).

The high incidence for the period 1200–1400 is of course due to the very marked Romance influence at the time: most of the new names in the period are of Latin/French origin.

---

[10] One exception is *lavender* (cf. Bierbaumer 1979, p. 156), borrowed directly from the French as its form reveals (Lat. *lavendula*). Another is possibly *perwince* from French *pervenche* (Lat. /*vinca*/*pervinca*). *Periwinkle* (Turner 1538 *perwyncle*) seems to be an early 16th century formation (Grigson 1974, p. 167).

[11] Cf. the French *langue de boeuf* commonly used in ME and 16th century herbals (incl. GH).

[12] The percentage of ME plant names in GH is probably higher than 55 % since some of the names first attested in GH, though not evidenced (so far) in ME records, are presumably of ME heritage.

[13] These names are: agaric, alleluia, apple of Paradise, archangel, basil, betony, buck's horn, calf's foot, cinnamon, cuckoo-pintle, five-leaved grass, gall nut, [hops], lady's seal, mushroom, nut of India, oak fern, paigle. For *hops*, see MED (1440 for 'Humulus') and cf. Bierbaumer 1979, p. 140.

[14] A few names recorded in ME and GH have OE "correspondences" in the form of direct reflexes of the Latin word, e.g. *fic* from Lat. *ficus*, where ME/GH *fig* (OF *fig(u)e*) is not a continuation of the OE form. Other such OE records of GH plant words are *betonice* (Lat. *vettonica*/Med. Lat. *betonica*), *lactuce* (Lat. *lactuca*) and *petersilie* (Lat. *petroselinum*). Cf. also OE *rŭde* (Lat. *ruta*). For the corresponding GH forms, see the Overall list.

Of the GH names of ME heritage only some 55 or 20% consist of Germanic word elements (cf. above for the different situation with the GH names of OE provenance). These names are: arsmart, black ivy, bloodwort, buck's horn, calf's foot, cleavers (OE clife), crowfoot, crowsoap, cuckoo-pint(el), dead nettle, dodder, dove's foot, dropwort, duck's meat, dwale, five-leaved grass, flag, goat's leaf, hart's tongue, honeysuckle, henbane, hops (cf. above), houseleek, knotwort, lichwale (see Brodin 1950, p. 225), lichwort, liverwort, maidenhair, mare's tail, mayweed, mouse-ear, oak fern, paigle (cf. note in the Overall list), pennywort, reed, red briar ('Rosa canina'), red dock, red madder, ribwort (OE ribbe), ruds, scabwort, self-heal, sowthistle, sparrow tongue, thistle of the sea, three-leaved grass, toadstool, wartwort, wolf thistle, woodcrab, wormwood and a few names in *wild* or *garden/tame+* noun (see the Overall list). For knee holme, see above.

A good many names are Germanic/Romance "hybrids" like *churl's treacle, devil's bit, horsemint, shepherd's purse, smallage* (small + OF *ache* < Lat. *apium*), *swinefennel* and *St. John's wort*. Some of these are of the type *wild* + non-Germanic word like *wild borage, wild mallow* and *wild sage*.

As for the GH names recorded before 1526 and with correspondences in the French text, it is difficult to ascertain whether the writer of the *Herball* took them straight from *Le grant herbier* or knew them beforehand. But, anyhow, the French text "strengthened" his knowledge here. Some examples in point are *hart's tongue* (Lat. *lingua cervi*), *lady's seal* (Lat. *sigillum sancte marie*), *mare's tail* (Lat. *cauda equina*) and *nut of India* (Lat. *nux indica*). Cf. above for similar names of OE provenance. Such names may consist of Germanic elements like *goat's leaf* (Lat. *caprifolium*) and *hart's tongue* or may be "hybrids" like *lady's seal*.

C. NAMES FIRST ATTESTED IN THE HERBALL

Overall list

The following English plant names are first (or only) recorded in *The Grete Herball* (1526).[1] Items not supplied with a reference (OED and/or Britten & Holland) are my own first-record discoveries. For additions and antedatings to the OED, see Chapter 5.

affodylly (word form only), agriot, artetyke (Britten & Holland), bear's

---

[1] In this list (and the following) the names are given in modern form (if any). For GH forms, see the Overall list.

The 1561 ed. of GH supplies the following (additional) first or unique records: garleke/water germander (Teucrium scordium), the lesse housleeke (Sedum acre?), waterwort (Crithmum maritimum; earlier for other plants).

bough, bear's foot, bear's twig, bistort, black bryony, blue flower-de-luce, bright (Britten & Holland), camel's straw, carpenter's grass (OED/Britten & Holland), case-weed, chervell 'Lonicera' (Britten & Holland), chibol of the sea, citron, (cf. note in the Overall list), coronary, cowgourde, crayfery (Britten & Holland), croyt marine, cuckoo's bread (OED/Britten & Holland), cuckoo's meat (OED/Britten & Holland), damask plum, date of India, dog rose, elf-dock (Britten & Holland), earth-thought, frog's foot (OED/Britten & Holland), fuller's grass (OED/Britten & Holland), fume of the earth (Britten & Holland), gall nut, gangelon, garden ginger, garden mallow, goosebill (Britten & Holland), goosefoot, Alexander's gourd, grass of the field, great bur, great centaury, great/small mugwort, great plantain (OED), greens (OED/Britten & Holland), hare trefle, hare's ballocks, hare's beard (Britten & Holland), hare's lettuce, hare's palace (Britten & Holland), herb of India, herb of musk, herb (grass) of wine (OED/Britten & Holland), [herb] paralysy (OED/Britten & Holland), herb Philip, herb squynantyke (OED/Britten & Holland), Hercules' grass, he-fern, hig(h) taper (Britten & Holland), high mallow, hog's meat, holy thistle, king's crown, knotgrass, lentils of the water, less bur, less mugwort, less saxifrage (OED), less skirret, licegrass, lingwort, little bur, little clote, little plantain, long plantain (OED/Britten & Holland), Margetym gentyll (the form), mederacle (the form), meu, middle mugwort, nespyte, oleander, orient saffron, parsley of Macedonia, polytrich, pompion, priest's hood (OED/Britten & Holland), pyllulary, quince apple, remcope, St. Peter's wort (OED/Britten & Holland), saracen's mint (OED), sea onion, senacion (OED), smoke of the earth (Britten & Holland), snakegrass, Solomon's seal (Britten & Holland), spewing nut, spike Celtic, sugar reed, strofulary, swine bread, tamaryte (the form), tapsebarbe (OED), three-cornered rush, tintymall (the form), tintymall of Babylon, wallfern (OED/Britten & Holland), water flag, water onion, white bryony, white vine, wild aloe, wild apple, wild blackberries, wild galingale, wild leek, wild mulberries, wild rape, wild smallage, wild valerian, wilding (OED), woodnep (OED), woodsorrel (OED/Britten & Holland), woodyp, yellow flag.

In addition, there are a few "names" with *common* or *tame* which seem to be first attested in the *Herball*, as *common garlick/onion/smallage* and *tame cress/onion* (in most cases reflexes of French designations in *commun(ne)* or *jardin* like *commune ache* and *cresson de iardin*).

Of the 124 names listed above (names in *common/tame* not included)—41 of which are part of chapter headings—91 or about 18% of the total of the English plant names mentioned in the *Herball* are noted here for the first time as GH first and/or unique records (cf. Chapter 5). Presumably, some names are of Middle English provenance, but are not in the published volumes of the

MED or (to my knowledge) in the citation files of the MED collections. Two examples in point are *cuckoo's meat* and *knotgrass* (recorded in ME as *knotting-grass*) which were known to Turner in 1538 and included in Turner 1548, in the table "for the commune english names vsed nowe in al countreis [i.e. counties] of Englande". The list of names given there is probably a rather haphazard one, but it is noteworthy that, for instance, neither of his other names for *Oxalis acetosella*, viz. *alleluia* and *woodsorrel*, are included, which may indicate varying frequency in popular usage. It is also noteworthy that *cuckoo's bread*, which occurs in the *Herball* for *Oxalis acetosella* as a mere translation of the French name *pain de cocu* (Lat. *panis cuculi*), is not mentioned by Turner.[2]

To make out, at least tentatively, which of the GH first-record names occur independently of the French text and which are translations, adaptations or adoptions of the French/Latin names as given in *Le grant herbier*, the English and the French texts have been collated. The result of this comparison is presented below.

Translations/adaptations/adoptions of names in the French text

The majority of the GH first records (some 80 or 63 %) are translations (total or partial) of French/Latin names as appearing in *Le grant herbier*. Such textual parallelism does not exclude independent knowledge of a name on the part of the translator (cf. pp. 13 and 35). It only implies that the French text was instrumental in the translator's name-giving. The French/Latin names are given here as appearing in the French text used.

Translations

bear's bough (foot, twig)	Brancha vrsina/branche vrsine
black/white bryony	la noire/la blanche brione
bright[3]	esclere
camel's straw[4]	paille aux chameaulx
carpenter's grass	herbe aux charpentiers
chibol of the sea	cibole marine
cuckoo's bread	pain de cocu
damask plum	prunes de damas
date of India	dates indes
dog rose	rose au chien/rose canine[5]

---

[2] There are corresponding names in other languages (see e.g. Lange, Lyttkens and Marzell).
[3] *Bright* as plant name now only in the compounds *bright-eye*, Ranunculus ficaria (Wright), and *eyebright*, Euphrasia; *eyebright* is in GH 1561 (earliest OED record from 1533).
[4] Cf. the Overall list and ME *camelys chaffe* (MED c. 1425) for the same plant.
[5] "une maniere de champignon". Cf. Fischer 1929 (p. 281): "Rosengallen". *Dog rose* 'Rosa canina' is not evidenced until 1597 (OED).

earth-thought[6]	
frog's foot	cf. ranoullie (< Lat. ranunculus 'small frog') in the text
fuller's grass	herba fullonum/herbe a foulon
fume of the earth	fumus terrae/fumeterre
gall nut	noix de galle
garden ginger	gingembre de iardin
garden mallows	Malua ortensis/mauue(s) de iardins
goosebill	bec doye
goosefoot	pied doyson
Alexander's gourd	courge de alexandrie
grass of the field	herbe des champs
great bur	la grant lappe
great centaury	la grande centoire
great mugwort	la grande armoise
great plantain	le grant plantain
hare trefle	trefle au fieure
hare's ballocks	cf. testiculis leporis in the text
hare's lettuce	Lactuca leporina/laitue a fieure
hare's palace	Palacium leporis/le palais au fieure
herb of India	herbe indaique
herb of musk	herbe de musc
herb (grass) of vine	herbe de vigne
Hercules' grass	herbe dercules
he-fern	Filex masculus/feugiere masle
high mallow	haute mauue
hog's meat	pain a porc
holy thistle	chardon benyt
king's crown	coronne royalle
lentils of the water	lentille deaue
less bur	lape la mendre
less mugwort	armoise la mendre
less saxifrage	petite saxifrage
less skirret	eruque petite
licegrass	herbe aux poulx
little bur	lape la mendre
little plantain	petit plantain
middle mugwort	armoise moyenne
orient saffron	saffren dorient
parsley of Macedonia	Petrocillum Macedonicum
priest's hood[7]	

[6] The *Arbolayre* and the "1500" ed. of *Le grant herbier* have *sourcir de la terre* (for *soucil de la terre* < Lat. (*Circa instans*) *supercilium terre* 'Adiantum', a fern). Cf. Huguet 7, p, 61, and Marzell 1 (1943), 490. The 1513/1540 and 1545 eds. have, however, *soucie de la terre* and *soulcie de la terre*, resp., forms which may occur in other, early 16th century, eds. of *Le grant herbier* (presumably available to the translator) and which may be the source of *thought* in the English text (cf. OED thought 5). For the use of 16th century eds. of *Le grant herbier* by the English translator, cf. *damacenes* in the Overall list.

[7] Cf. the French text: "Aucuns lappellent vtila preste [? prestre] car elle a vne telle chappe" (as in the other eds. seen).

pyllulary[8]	
quince apple	pommes de coing
St. Peter's wort	herbe saint pierre
saracen's mint	cf. sarrazine/mente romaine
sea/water onion	cepa marina/oignon marine
small mugwort	[see sub less mugwort]
smoke of the earth	fumus terre/fumeterre
snakegrass	De serpentina/serpentine
Solomon's seal	sigillum salomonis/le sel solomon
spewing nut	noix vomique
sugar reed	cf. canna mellis[9]
swine bread	pain a porc
three-cornered rush	ionc a trois costes
tintimal of Babylon	tintimal babilonique
white vine	vitis alba/vigne blanche
wild aloe	aloes sauuages
wild apple	pomes sauuages
wild leek	porreau sauuage
wild mulberries	moures sauuages
wild rape	raue sauuage
wild smallage	ache sauuage
wild valerian	valeriaine sauuage
water flag	glay de eaue

In some cases, the wording of the French text may have been influential, as with *wallfern*: "elle croist contre les murs". Cf. also above, notes 7 and 9.

Adaptations

Of the remaining names, three represent a type of modication of French/Latin names well evidenced in Middle English:[10] *affodylly* (< *affodile*), *coronary* (< *coronaire/Coronaria*) and *strofulary* (< *strofulaire/Strofularia*). *Chervell* is a corruption of French *chèvre-feuille* (Lat. *caprifolium*). *Cowgourde* may be the French *cougo(u)rde/cowgurde* (cf. Grigson 1974, p. 90) or it may represent a substitution of elements, viz. of *gourde* for *courde* as part of the *coucourde(s)* of the French text. *Senacion* represents *senation* (< Med. Lat. *senacionem*) in the French text.

---

[8] Used as a synonym for *licegrass* (see above). It renders the French *pediculaire* (< Lat. *pedis/pediculus* 'louse'). Cf. Brodin 1950, pp. 246 f.

[9] Cf. the French text: "Canna mellis/cest la plante ou le succre croist."

[10] Cf. GH names like *centaury, eupatory, policary, polypody, sanguinary* and *saponary*, all attested in ME.

Adoptions

Of the GH first records some 18 are adoptions of French names, such as *agriot*, *artetyke*, *bistort*, *croyt marine*,[11] *herb squynantyke*, *herb paralysy*, *herb Philip* (< *herbe saint philippe*), *nespyte*, *pompion*, *meu*[12] and *tapsebarbe*. For the categorization of such names as "English", see p. 31.

Other names

Of the plant names which have their earliest attestation in *The Grete Herball* the following are not textual parallels, i.e. they are given independently of the French text:

blue flower-de-luce, case-weed, citron, crayfery, cuckoo's meat, elf-dock, gangelon, greens, hare's beard, hig(h) taper, knotgrass, lingwort,[13] long plantain (in the French text *petit plantain*), Margetym gentyll,[14] mederacle, remcope, wild blackberries, wild galingale, wilding, woodnep, woodyp, woodsorrel, yellow flag.

It should be noted that most of these are compound or phrasal. Only one (of the clearly identifiable names) is a simplex (*greens*) and one (*wilding*) a derivative. For *crayfery*, *gangelon*, *mederacle*, *remcope* and *woodyp*, see below. On the whole, less than 10 % of the English plant names first attested in the *Herball* are non-compound or non-phrasal, a distribution chiefly due to the great element of translations among the GH first or unique records. Compare here Barber 1976 (p. 191), who states that "a very large group [of new compounds] is formed by the names of trees, plants, and birds, especially plants ... These are especially common in the 16th century, and tend to be popular names, or the names used in Herbals, rather than scientific ones".

Dubious names

The 'origin' of the English plant names in *The Grete Herball* can in most cases be established with reference to earlier and later occurrences in English records and/or to plant names as found in *Le grant herbier*. There are however

---

[11] Recorded as *crest-marine* in OED (1565) and in Lyte (1578) and Gerard (1597). In GH 1561 *waterwort* is given as an equivalent of *croyt marine* (cf. above).

[12] Cf. Britten & Holland (p. 333): "invented by Turner (Names)".

[13] "Lyngewort, or peleter of Spayne".

[14] Or *gentyll Margetyn* (for *Origanum majorana*). The name shows great formal variation. Other GH forms are *Magerym/Margerym/Margarym*. Banckes's herbal (1525) has *Magerum*, Turner 1538 *Margerum/Margerum gentyll*, Turner 1548 *Margerum/Mergerum*, Turner 1568 *Marierum gentle*. For ME forms, see MED *majorane*. The element *tym* (*tyn*) in the GH name possibly reflects *thyme* (by way of popular etymology).

some "first records" in the *Herball* – with no equivalents in *Le grant herbier*, at least in the editions available for the present study—which present problems of provenance and derivation. These are:

crayfery (De pulmonaria), gangelon (De satirione), remcope (Morsus diaboli), woodyp (De genestula) and mederacle (De camephiteos)

It should be observed that the names are all part of chapter headings and are also included in the "Registre". They also occur in the other editions of the *Herball*.

crayfery may be a composition of *cray* 'crop of a bird' (OED craye 'craw') and *fer(r)y* 'bread'. The name refers to a lichen (as represented in the picture)—probably *Lobaria pulmonaria*—though the text supplies a description of the lungwort, *Pulmonaria officinalis* (cf. Britten & Holland, p. 127). In the 1561 edition the word is given as *crafery*, where *cra* may represent a dialect form (cf. Wright 1 (1898). p. 780). Lichens were formerly used, in times of famine, for bread-making.

gangelon, which is used in the *Herball* as a synonym for *hare's ballocks* and *satyrion*, refers to an orchid. The word may be a modification of *ganglion* 'swelling' alluding to the root tubers of some orchids.

remcope (in the "Registre" *remcop*) may contain French *ren* 'again' (*ren* >*rem* by dissimilation) and *couper* 'bite'. Cf. also French *remordre* 'bite again'. The name refers to *Succisa pratensis*, for which *Morsus diaboli* and *devil's bit* are used as synonyms in the *Herball*. For these latter names, see Grigson 1974, p. 70.

woodyp contains *wood* and possibly *hip* (cf. Earle 1880, p. civ, and MED hepe n. 2), with loss of the initial consonant as part of the compound. The botanical reference of the word remains however doubtful, though a small shrub of some kind seems to be intended.[15] In the *Herball*, where there are no synonyms for *woodyp* either in the text or in the "Registre", "De genestula" immediately precedes "De genesta" (i.e. broom).

None of the four plant names dealt with above are recorded in the OED or the MED.

Mederacle, which appears sub "De camephiteos" (as part of the chapter heading), "De camedrios" and "De cameleonta" ("camephiteos that is mederacle"), is probably identical with *med(e)ratele*, i.e. mead-rattle, meadow-

---

[15] The text runs as follows: "Genestula is an herbe lyke to brome/but it is lesse and hath smaller braunches and twygges and hath a whyte floure/and a reed sede as brust or fragon/or kneholme [i.e. *Ruscus aculeatus*] whiche be allone but genesta hath a yelowe floure". In Banckes's herbal "Genestula" is broom: "This herb men call it Genestre or Brome". Camus 1886 (p. 71) identifies the plant as *Osyris alba*.

rattle, *Rhinanthus* (cf. Britten & Holland, p. 330), evidenced in mid-15th century English sources (see MED mederatele, OED mead-rattle and Brodin 1950, p. 242). The GH form may be due to misprinting of *c* for *t*—confusion of these letters is common in ME and early modern English manu-scripts[16]—or to the translator being influenced by the French name *quercule*, in GH appearing as *quercle*.

Cf. the French and English texts (sub "De Camedreos" and "De came-drios", respectively):

"Cest vne herbe quon appelle gemandrea. et est autrement appellee carcula-minor/la mēdre quercule. Et la grāde quercule est camephiteos et est vne mesme herbe. Camedreos la mendre quercule est gemandree."

"Camedryos . . . it is an herbe called Germaūdrea or quercula minor ye lesse quercle the grete quercle is called camephiteos that is mederacle/Camedrios is the lesse quercle and germaūdre." As in *Alphita* (c. 1450), *mederacle* is here given as a synonym for *quercula maior*, as opposed to *quercula minor*, a germander. Cf. Brodin 1950, *loc. cit.*

D. VARIATION IN FORM AND SPELLING

As is well known, there is in early documents—in the absence of an official norm—considerable graphemic/morphological variation. Plant names, as part of an oral rather than a written tradition, have always tended to vacillate both in form and 'meaning' (i.e. botanical reference). In fact, formal and semantic instability is a marked characteristic of plant names, especially of names in popular use. And the vagaries of folk etymology are often mirrored in plant names.[1]

A distinction must of course be made between mere graphemic variation, such as *i/y* variation or variation in the use of final *e*, and genuine phonemic/morphological variation.

Examples of graphemic variation, which is very common in the *Herball*, are *an(n)es/an(n)ys*, *barl(e)y/barli*,[2] *borage/bourage* (cf. below), *cockle/cocle/cock-yll/cokyll*, *gourd(e)/gowrde*, *gromell/gromyll*, *hemlocke/hemloke*, *letuce/letuse*, *nettel(l)/nettle*, *plantain/plantayn(e)*, *rew/rue*, *rewbarbe/ruberbe*, *sinkfoyle/synkefoyle*, *wyloue/wilowe*, etc.

Some names show variation which is due to phonetic interference, as in *alysander/alysamder*, *cynamonne/cyn(n)amome*, *tintymall/tintinall*, *henbane/*

---

[16] Cf. also Bergh 1978 (p. 8): ". . . the similarity between *t* and *c*, easily observed in Medieval Latin manuscripts."

[1] Two exx. are *wormwood* and *meadowsweet* (see Grigson 1974).

[2] Cf. Dobson II (1968), p. 845. ME forms are *barli*, *barlich/barlick* and *berley* (OE *bærlic*). See MED.

hembane/henbame,[3] *dytany/dyptan(y)*.[4] Cf. Dobson II (1968), pp. 954 and 1002. In *strammarche* (for *stammarche*)—sub "De Petrecilio macedonico"—there is an 'excrescent' *r* (the form is not in the other eds. of GH and may simply be a misprint). For *comfrey/confrey*, cf. Wyld 1953, p. 294, and MED conferie (OF *confirie*/Med. Lat. *cumfiria*).

For variations of the type *duck meate/duckes meate*, see the Overall list.

There are not a few cases where the variation is conditioned by historical/ etymological factors, i.e. by the immediate sources of the respective forms. Some examples:

agrymony (Lat. agrimonia)	egrymony (OF aigremoine)
alexandre/alexandry (Med. Lat. alexandrum/OE alexandre)	alysander (OF alisa(u)ndre)
bo(u)rage (Med. Lat. bor(r)ago)	bo(u)rache (OF bourrache)
calament (Med. Lat. calamentum/ OF calemente)	calamynt[5]
eufragye (Med. Lat. eufragia)	eufrace (Med. Lat. euphrasia/ OF eufraise)
fen(e)greke (OF fengrec)	fenugrec/fenygrec (OF fenugrec)[6]
mercury (Lat. mercurius)	mercuryall (OF mercuriel)
myllefoly (Lat. milifolium/ millefolium)	myllefoyle (OF milfoil/milfueil)
peleter/pellyter (Lat. pyrethrum)[7]	pireter/perytory (Lat. parietaria)[7]
pimpinell (Med. Lat. bipinella/ pipinella)	pympernell (OF pimprenele/ pimpernelle)
portulax (Lat. portulaca)	po(u)rcelaine/purcelane (OF po(u)rcelaine)

In one case, *chervel(l)*, one form reflects different origins, viz. Lat. *caerefolium* (OE *cerfille*) 'Anthriscus cerefolium' and French *chèvre-feuille* (Lat. *caprifolium*) 'Lonicera periclymenum'.

Occasionally, a name misprinted in the 1526 edition of the *Herball* is corrected in a later edition. One example in point is *sperworde/sperewort* (1561) for *sereworde/sereworte* (1526) sub "De flamula".[8] For *crayfery/crafery*, see p. 41.

---

[3] There is also a form *hanebane* (sub "De frumento"). Cf. Brodin 1950, p. 229. Cf. also *henlocke* (for *hemlocke*) in VBD.

[4] Cf. also *rampsons* (for PE *ramsons*) and *solempnell* (sub "De eufragia": "Mayster Peter of Spayne that was a solempnell clerke").

[5] Cf. MED calaminte.

[6] In ME the word shows great formal variability (see MED).

[7] The corresponding OF forms are *peletre/piretre* and *paritaire*. The current form *pellitory*, which is due to suffix change and/or *r/l* dissimilation (see OED pellitory and Larkey & Pyles 1941, p. 28), dates from the mid-16th century.

[8] Cf. Banckes's herbal (1525) sub "Flamula": "Flamula is called Spereworte or Launcell". See also Britten & Holland, p. 445. *Sereworde*, which is also in the eds. of 1529 and 1539, may possibly be a genuine form, containing *sere(sear)* 'burn, scorch' referring to the plant as causing sores and blisters. Cf. Sw. *brännört* (for *Ranunculus sceleratus*).

3. Plant-name explanation in the Herball

As is customary in medieval herbals, many plant names mentioned are 'explained' in *The Grete Herball*. These explanations—usually of transparent, self-explanatory (mostly compound) names—transmit traditional knowledge, which was part of popular and/or learned herb knowledge. In fact, such etymologies "were the kind of reasoning that ensured the popularity of the herbal" (Stannard 1974, p. 30). They mirror a genuine interest in word origin, particularly of course in the connection between the name and the use or a characteristic feature of the plant in question. They also imply a good deal of folk lore. In most cases, the translator closely follows the French text. A few citations:

It is called Agnus castus/chaste lambe/for it kepeth a man chast as a lambe/and withdraweth lechery

Bursa pastoris ... is shepeherds purs some call it sanguynary bycause it stauncheth the bledynge of the nose ... the sede of it is lyke a purs

Calendula ... is called calendula bycause it bereth floures all the kalendes of euery month of the yere

Candelaria [*Verbascum*] is an herbe that is so named bycause it is like a tapre of waxe

Centaurium ... is a veray bytter herbe/and therfore it is called erthe galle

yᵉ deuylles bytte ... is so called by cause the rote is blacke and semeth that it is iagged with bytynge/and some say that the deuyll had enuy at the vertue therof and bete the rote so for to haue destroyed it

One is called whyte elebore bycause the rote is whyte/and bycause it purgeth white humours/as flewmes

Flammula is an herbe so named bicause it is hote and brenneth as flamme

Fumus terre ... bycause it cometh out of the erthe in grete quantyte lyke smoke

Goosbyll ... The rote of it is lyke a goos byll

Goos fote bycause the sede spredeth forkewyse as a goos fote

It is called hares palays. For yf the hare come vnder it/he is sure that no best can touche hym[1]

Penthafilon is an herbe called fyue leued. For pentha in greke is .v. and filo is leef. And so penthafilon is to say herbe with .v. leues

Pilocella or mows eare ... hath rough leues wᵗ longe heares in them lyke a mous eare/and therfore is so named

Primula veris ... is called prymerolle or primula/of pryme tyme/bycause it bereth the fyrst floure in pryme tyme

---

[1] In the French work the picture includes a hare hiding under the leaves of the plant.

Rewbarbe . . . groweth in Inde or barbary/and therfore it is called rewbarbe. The other is rewponticum/bycause it groweth in an yle called ponticum

Saxifrage is so called bycause it breketh the stone

Serpentina is otherwyse called dragons/or snakegrasse bycause the stalke is spekled lyke a snake

Squinant is an herbe that is called camelles strawe. bycause camelles do eate it

Herbe rabious that some cal wartwort/bycause it is good for wartes or ryngwormes.

4. *Problems of identification and plant-name equivalence*

The study of plant names in early documents is beset with many difficulties, the most salient being that of identification, i.e. the ascertainment of the relation between the 'name' of a plant and its referent (i.e. the object it represents).[1] This applies in particular to the medieval and early 16th century herbals with their inadequate descriptions and poor (if any) illustrations which in most cases are not true to nature and which are often misapplied.[2] In the analysis of plant-name reference and plant-name equivalence in *The Grete Herball* other aids must often be resorted to, notably the MED, the OED, Britten & Holland 1878–86, Earle 1880, Camus 1886, Fischer 1929,[3] Marzell 1943–79, Brodin 1950, the 1941-edition of Banckes's herbal (1525) and the 1965-edition of Turner's *Libellus* (1538) and *Names of Herbes* (1548). However, even with the assistance of these works (and others) doubt often remains as to the precise identity of a plant as mentioned in the *Herball* and in the Overall list I have in some cases refrained from giving a definite assignment of a name to a particular species rather than doing so for the sake of a 'complete' analysis. On the other hand, the plants mentioned in the medieval herbals tend, at least to some extent, to be the same, which of course facilitates the

---

[1] For the definition of *plant name*, see above, p. 31. A "plant name" basically refers to a species (or another taxon) but it may also, in a given situation, indicate a particular specimen of a species (subspecies, etc.). On the problem of identifying early plant names, see e.g. Earle 1880, pp. lxi ff., and Singer 1961, pp. xl ff.

[2] Cf. Prior 1863 (p. ix): "The Grete Herball, the Little Herbals, and Macer's Herbal . . . and some other black-letter books of an earlier date than Turner's, are of scarcely any assistance to us, from the difficulty there is to discover by their inadequate descriptions, what plants they mean." As hinted by Prior, the situation is largely different in the herbals by Turner, Lyte and Gerard, provided as they are with more accurate woodcuts taken from Fuchs and Tabernaemontanus (Gerard). The situation is of course also different in the case of the early, scholarly floras.

[3] See especially the section in Fischer entitled "Die im abendländischen Mittelalter in der Literatur genannten Pflanzen. Zugleich: Synonymenschlüssel". The identifications given there are however not always applicable to GH.

identification of the plant names in the *Herball*. And the plants described in the old herbals are usually the well-known ones (cf. Introduction 4). As pointed out by Singer 1961 (p. xliv), "rare species were not noticed even by such an ancient master of botanical knowledge as Theophrastus; still less would they be known to a barbarian scribe. Thus most of our flora may be safely disregarded in any attempt to identify ancient plant-names".

There are however other difficulties in the identification of early plant names. One is the considerable referential instability of plant names—a plant name may be synchronically applied to more than one species and its reference may change with time—another is the tendency in early botanical works to transfer vernacular names to non-native plants, not distinguishing the native flora from non-native ones (cf. p. 24). The enumeration of foreign "synonymous" plant names in (certain) old herbals (see p. 30) is only of marginal avail to a modern reader.

William Turner considered *The Grete Herball* to be "al full of vnlearned cacographees and falselye naminge of herbes" (The Preface, Turner 1568). One of the reasons for this "false naming" is that, as stated in the Introduction, the *Herball* is largely a translation of a French herbal (*Le grant herbier*), which in its turn is, in part, a reflex of a Latin work (*Circa instans*). This series of textual dependencies causes problems for the identification of the English plant names as occurring in the *Herball*, chiefly in terms of discrepancies between the current, i.e. early 16th century, botanical reference of the English names[4] and the botanical reference as indicated in the text and in the Latin chapter headings (as overtaken from the Latin/French sources). In other words, the botanical reference of a name as mentioned in the *Circa instans* and/or *Le grant herbier*,[5] mirroring floras partially differing from that of England, is not always identical with the botanical reference as indicated by the "parallel" English name provided by the English translator. Such referential anomalies are commented on in the Notes to the Overall list. The discrepancy may be one between related plants, as with *Oxalis corniculata/O. acetosella* ("De Alleluya") or *Arum italicum/A. maculatum* ("De Iaro") or one between unrelated plants, as exemplified sub, for example, "Ameos", "Consolida media", "De palacio leporis" and "De silfu" (cf. the Overall list).

Crucially linked with the problem of identification is the problem of plant-name equivalence, i.e. the establishment of the referential identity of plant names. Obviously, we cannot here be more exact than the authors of the herbals were. Our primary task must be "to sift out the certain from the uncertain" (Earle 1880, p. lxvi).

---

[4] As can be inferred from e.g. OED quotations, Banckes's herbal of 1525 and Turner's works, in particular his *Names* of 1548.
[5] As identified by Camus 1886, Wölfel 1939 and others.

Plant names, like other terminologies, render extralinguistically established distinctions; they refer to different taxa of plants (species, subspecies, etc.). But they may also imply "sameness"; they may have referential identity. Plant names with referential identity[6] are generally said to be synonyms. Such names, for instance *cuckoo's meat* and *woodsorrel*, have the same referent, i.e. a specific object (a plant taxon) in the external world, as subsumed under a scientific (Latin) formula, in the current nomenclature *Oxalis acetosella*. But coreferential plant names are not part of a common conceptual sphere—they are not realizations of any identical cognitive (descriptive) meaning—though they may be stylistically/situationally differentiated, i.e. they may differ in "emotive meaning" (see Fries 1980, p. 35, and cf. Lyons 1977 (I), p. 175; on plant names as contextual elements, as part of literary texts, see Rydén 1978 a). For plant names, as for other lexical items that have reference, identical reference is, in other words, "a necessary, but not sufficient, condition of synonymy" (Lyons 1968, p. 427). And coreferential plant names in a herbal or flora are only cotextual synonyms, as collected (or translated/invented) by the author of that particular herbal or flora. Hence a term like *equivalent*, as being less loaded than *synonym*, seems more appropriate here (cf. Fries 1980, p. 32).

For ascertaining *plant-name coreferentiality* in *The Grete Herball* there are at least four possible means, three cotextual and one extratextual:

1. (*a*) Co-occurrence of names in chapter headings
   (*b*) Co-occurrence of names in heading and immediate context (i.e. the text appended to the heading)
   (*c*) Other parallelizations of names as given in the text of GH
2. Name equivalence as inferred from texts other than GH (contemporary or non-contemporary with GH), such as herbals, floras, dictionaries, etc.

For exemplification of (1 a) and (1 b), see above sub "Arrangement of the plant names". Some examples of (1 c) are:

iuce of arssmert or persicaria (De herba indica)
colloquintida or wylde gourdes (Anacardus latine & grece)
eupatorium/or wylde sawge (De Ambrosiana)
fenugrec or setwal (De Altea)
eruca or skyrwit the lesse (De bursa pastoris)
poligonia that is knotwort or swynegrasse (Coronaria)
sanguinary that is bursa pastoris/or cassewede (De bolo armenico)
scaryole that is wylde letuse (De Boragine)
Baucia is an herbe called skyrwyt ... It is all [so] called pastinaca (De baucia)

---

[6] Not to be confused with identical referential (descriptive/cognitive) meaning (see below).

And whan cresses is onely spoken of without ony addycyon it is gardyn cresses (De Narsturcio)

None of these methods is of course entirely water-proof, though (1) as cotextual should, at least in principle, be more reliable than (2). Naturally, the establishment of plant-name coreferentiality does not necessarily imply plant identification.

Pertinent to the discussion of plant-name variation and equivalence is the question which plants have many names and which have few or no (non-scientific) names.[7] Although the *Herball* is comparatively rich in coreferential plant names (see below), the name-giving there does not allow us to draw any far-reaching conclusions as regards plant-name distribution. However, it seems obvious that, generally speaking, widely spread plants and easily recognizable plants have more (popular) names than rare plants and than plants which are difficult to recognize (and distinguish) like grasses and sedges.[8] Further, plants that attract attention in some way or other, positively and negatively, are given many names, i.e. plants that stir people's imagination (like *Arum* and *Oxalis*) or plants that are considered useful (like *Achillea* and *Saponaria*) or troublesome plants, for example persistent weeds like *Agrostemma* (now increasingly scarce), *Agropyron repens*, *Arctium*, *Galium aparine* and *Polygonum aviculare*. As noted above, there is a correlation between the commonness of a plant and the number of (popular) names assigned to it, but this does not imply that all common and well-known plants have many popular names. Certain plants such as some trees, however well known, useful or spectacular, have few names; commonness is, in other words, no guarantee for many names; some other factor, factual or non-factual, is needed.

A question relevant to the discussion of the coreferentiality of plant names in the *Herball* is whether the arrangement of the (coreferential) English names in the headings and in the accompanying texts can tell us anything about their relative frequency at the time. Although the sequence of the English names in the text appears to be largely haphazard, except that a translation of a French name usually occurs in its "proper" place in the English text, the names as given in the chapter headings probably represent, in the great majority of cases, the more common names of the day, at least in the translator's view.[9]

Given the difficulties of identification and species-delimitation outlined

---

[7] Cf. Fries 1980, pp. 32 f. Fries states that "important plants have few names", i.e. plants which have/have had a central position in man's life over large areas, e.g. certain trees and cereals.

[8] In the old herbals (prior to the mid-16th century), grasses and sedges are, on the whole, rarely mentioned; in e.g. Turner 1548 there are "no sedges, and only five grasses" (Turner 1965, p. 6). And even in the later herbals "grasses, rushes, and sedges were not regarded as really distinct groups" (Arber 1941, p. 29).

[9] Cf. p. 30.

above, the spread of plant-name equivalence as regards the English names in *The Grete Herball* (1526) is as follows (based on a total of some 300 species/genera):

plants with 1 name	59 %
plants with 2 names	23 %
plants with 3 or more names	18 %

Consequently, about 40 % of the plants mentioned in the *Herball* are assigned more than one English name. The highest number of such equivalents found is five (in one case possibly six).

The value of the *Herball* as a plant-name document is highlighted by a comparison with Turner's *Names* (1548), the only contemporary work for which a list of plant-name "synonyms" has been worked out (see Turner 1965). Here 43 plants, of the some 530 listed, or only some 8 % are given more than two names and only six plants[10] have more than three. In all, some 35 % of the plants mentioned in Turner 1548 have more than one name. The relatively high degree of plant-name synonymy in the *Herball* is partially due to synonyms taken over from the French original. In Gerard 1597 the highest number of English equivalents given seems to be nine (Rydén 1978 b).

In the lists below, the arrangement is alphabetical (in the sequence of equivalent frequency) according to the modern Latin name.[11] In some cases only genus is indicated. For uncertain references, see the Overall list.

Plants with three or more English names in GH

Bryonia alba/dioica	bryony, white bryony, white vine, wild gourd, wild nep, wild vine (?)
Achillea millefolium	bloodwort, carpenter's grass, milfoil, sanguinary, yarrow
Althaea	garden (tame) mallow, hock, hollyhock; high mallow, wild mallow
Arctium	clivers, clote, great bur, less bur, little bur
Iris	blue flower-de-luce, flag (?), gladdon, water flag, yellow flag[12]
Oxalis acetosella	alleluia, cuckoo's bread, cuckoo's meat, sorrel de boys,[13] woodsorrel

[10] Althaea officinalis, Brassica oleracea, Hordeum vulgare, Matthiola incana, Phaseolus vulgaris, Raphanus sativus.

[11] Modern spelling and form. In some cases the GH form is given or added. It is noteworthy that the majority of the plants in the lists can be identified as wild or naturalized British plants.

[12] *Gladdon* and *water flag* refer to *Iris pseudacorus*, presumably also *yellow flag* (despite the Latin heading; see the Overall list).

[13] "alleluya/that is sorell de boys or cukowes mete" (sub "De Agarico"). Cf. OED sorrel 3 a. In Banckes's herbal (1525) there are the following equivalents: alleluya, woodsowre, wyld sowre (misprinted as wyndsowre; see Larkey & Pyles 1941, p. 8) and stubworte. The earliest OED record of *stubwort* is from 1541. The earliest attested English names for *Oxalis acetosella* seem to be *gēacessūre* and *þrilēfe* (Bierbaumer 1975–79).

Polygonum aviculare	centinode, knotgrass, knotwort, sparrow tongue, swine's grass
Saponaria officinalis	borith, crowsoap, fuller's grass, herb Philip, saponary
Agrostemma githago	cockle, darnel, drawk,[14] ray
Anchusa (or "similar" plants)	bugloss, langdebeef, oxtongue, wild borage[15]
Arum maculatum	aron, calf's foot, cuckoo pintle, priest's hood
Bellis perennis	bonewort(?), bruisewort, daisy, the less consoulde
Brassica/Sinapis	less skirret (?), senvy, wild coles, wild rapes[16]
Eryngium maritimum	red briar, teasel, thistle of the sea, yringe
Lemna minor	duck('s) meat, frog's foot, greens, lentils of the water
Polygonum hydropiper	arsmart, bloodwort, culrage, sanguinary
Primula veris	artetyke, cowslip, herb paralysy, paigle
Rubus fruticosus	blackberries, bramble, wild blackberries, wild mulberries[17]
Urginea maritima	chibol of the sea, sea onion, squill, water onion
Verbascum thapsus	hare's beard, hig(h) taper, mullein, tapsebarbe[18]
Anthemis cotula	maidenweed, maithen, the middle consoulde
Artemisia vulgaris	mother of herbs, motherwort, mugwort[19]
Asplenium	earth-thought, polytrich, wallfern
Capsella bursa-pastoris	case-weed, sanguinary, shepherd's purs
Citrullus colocynthis	colloquintide, Alexander's gourd, wild gourd
Crocus sativus	garden saffron, orient saffron, saffron
Dracunculus vulgaris	dragons, serpentine, snakegrass
Euphorbia	spurge, tin(n)timall (i.e. tithymall) /of Babylon/, wartwort(?)
Fumaria officinalis	fumyterry (i.e. fumitory), fume of the earth, smoke of the earth
Gentianella/Gentiana	baldmoney, felwort; gentian

---

[14] See Britten & Holland, p. 159, and the Overall list.
[15] The names may refer to several rough-leaved plants. Cf. e.g. Britten & Holland.
[16] Specific species of *Brassica* are *caule wortes* (*B. oleracea*) and *rapes* (*B. rapa*).
[17] Only *bramble* refers to the bush (the other words to the fruit).
[18] Cf. (sub "De scabiosa"): "Tapsebarbe is a maner of herbe called moleyne ... and is called wolues tayles in frensshe", as translated from "Tapsebarbe cest molaine ... on lappelle queue de leu". See also OED tapsebarbe.
[19] Specified as "great", "middle", "less" and "small" (as reflexes of the French text) with uncertain ref.

Helleborus niger	(black) hellebore, lion's foot, pedelyon
Inula helenium	elf-dock, horseheal, scabwort
Lamium (especially L. album)	archangel, blind nettle, dead nettle
Lithospermum officinale	gromwell, lichwale, lichwort
Lonicera periclymenum	chervell, goat's leaves, woodbind
Malus sylvestris	wild apple, wilding, woodcrab
Melilotus	honysuckle, king's crown, melilot
Mespilus germanica	medlar, mespile, open arse[20]
Orchis/Dactylorhiza/	
Platanthera	gangelon, hare's ballocks, satyrion[21]
Plantago lanceolata	little plantain, long plantain, ribwort
Plantago major	great plantain, plantain, waybred
Primula vulgaris	paralysy, primerolle, St. Peter's wort[22]
Prunus domestica	damson, damask plum, plum
Rosa	rose; briar, thorn[23]
Rumex	dock, red dock, sorrel
Sempervivum tectorum	houseleek, jōbarde,[24] sengreen
Senecio vulgaris	groundsel, senacion, senechon
Smyrnium olusatrum	alexanders, parsley of Macedonia, stanmarch
Solanum nigrum	less morel, nightshade, petymorell[25]
Trifolium	hare trefle, three-leaved grass, trefle
Teucrium scorodonia	eupatory, hindheal, wild sage
Veratrum album	lingwort, pelitory of Spain, (white) hellebore

Plants with two English names in GH

Acanthus mollis	bear's bough/twig, bear's foot
Alchemilla	lion's foot, pedelyon
Amaracus (Origanum) dictamnus	dittany, garden ginger
Allium sativum	churl's treacle, (tame) garlick
Allium ursinum	ramsons, wild garlick
Anethum graveolens	anet, dill
Apium graveolens	smallache, stanmarch
Aristolochia rotunda	meek galingale, smearwort
Asperula cynanchia	herb(grass) of vine, herb squynantyke
Asphodelus	affodylly, woodruff

[20] A fourth GH "variant" is *nefle*, which is however only an adoption of the French name as given in *Le grant herbier:* "Nespile be medlers or nefles" translating "Mespile ou nespila/Ce sont nefles". The "Registre" has "Nespilus mydlers or nefles".
[21] Another (Latin) equivalent in GH, as in most old herbals, is *palma christi:* "Palma christi is an herbe lyke satyrion". Cf. Brodin 1950, p. 240.
[22] "Primula veris is called prymerolles. Some call it saynt peterworte. Other paralisie". *Primrose* is not in GH, a name which in Banckes's herbal is applied to "Ligustrum".
[23] Cf. *brere* in the Overall list. *Eglantine* is *Rosa rubiginosa*.
[24] As in the French text ("iōbarde") for *jobarbe/joubarbe* < *barba jovis*. Cf. OED jubarb.
[25] For bot. ref. of *morel*, see the Overall list.

Atropa belladonna	dwale, more morel
Calamintha	calamint/of the mountain/, nespyte
Calendula officinalis	marygowles, ruds
Castanea sativa	chesteine, chestnut
Centaurium erythraea	centaury, earthgall
Ceterach officinarum	ceterach, saxifrage (?)
Cichorium endivia	endive, scaryole
Cinnamonus	canel, cinnamon
Cnicus benedictus	holy thistle, sowthistle(?)
Coronopus squamatus(?)	buck's horn, wartwort
Corylus maxima	avellan, filbert
Cucumis melo	melon, pompion
Cucumis sativus	cowgourde, cucumber
Cucurbita pepo	citrul, gourd
Cyclamen europaeum	hog's meat, swine bread
Cymbopogon schoenanthus	camel's straw, squinant
Cyperus longus	three-cornered rush, wild galingale
Delphinium staphisagria	licegrass, pyllulary
Digitaria sanguinalis	goosefoot, sanguinary
Geum urbanum	avens, geloffre
Hedera helix	ivy, black ivy
Hypericum perforatum	herb John, St. John's wort
Lactuca serriola	scaryole, wild lettuce(?)
Laurus nobilis	bays, laurel
Lepidium sativum	cress, garden (tame) cress (for *senacions*, see Overall list)
Linum usitatissimum	flax, line
Lolium temulentum	cockle, ray[26]
Majorana hortensis	Margetym gentyll, mariorayne (i.e. marjoram)
Mandragora officinarum	mandragora, mandrake
Melissa officinalis	balm, melisse
Mentha gentilis (?)	saracen's mint ("mynte romayne"), white mint
Mentha longifolia	horsemint, wild mint[27]
Mentha spicata	mint, garden (tame) mint
Nardostachys jatamansi	spike, spikenard
Nigella sativa	cockle, gith
Origanum majorana (see Majorana hortensis)	
Origanum vulgare (?)	brotherwort, organ
Panicum mileaceum	mile, millet
Papaver rhoeas	red poppy, wild poppy
Papaver somniferum	black poppy, white poppy
Parietaria diffusa	pireter (i.e. pellitory /of the wall/?), wallwort
Peucedanum officinale	dog fenel, swine fenel

---

[26] Possibly also *darnel*.
[27] "Ther is an other mynte and it is wylde and is called mentastre or horsmynte" (De menta).

Picea (or Pinus)	fir, sapin[28]
Pimpinella anisum	anise, sweet cumin
Pimpinella saxifraga	pimpernel, selfheal
Polygonatum multiflorum	our lady's seal, Solomon's seal
Polypodium vulgare	oak fern, polypody
Potentilla reptans	cinquefoil, five-leaved /grass/
Ranunculus ficaria (?)	bright, celandine
Ranunculus sceleratus	ache, crowfoot (for s[p]ereworde, see Overall list)
Rubia tinctorum	madder, warence
Ruta montana	rue of the field, wild rue
Stellaria holostea (?)	goosebill, stichwort
Succisa pratensis	devil's bit, remcope
Symphytum officinale	comfrey, the more consoulde
Tanacetum balsamita	cost, costmary
Thymus serpyllum (?)	pellitory, wild thyme
Trigonella foenum-graecum	fenugreek, setwall
[Sub "Ameos"]	pennywort, woodnep (see Overall list)
a lichen	crayfery, lungwort

As appears from the lists given above, the set of equivalents is usually a mixture of native and foreign names. The high proportion of non-native names is of course due to the textual background of the *Herball*. The variant ("synonymous") names may however be solely of native stock, as with *Inula helenium*.

Whereas about 40% of the plants (or fruits) in the *Herball* are supplied with more than one English name, one name is comparatively seldom applied to more than one plant (cf. Brodin 1950, pp. 208 f.). Some instances are given below (for further exx., see the Overall list). Botanical references antedating the earliest OED record or references not found in the OED are asterisked.

bloodwort	Achillea millefolium,* Polygonum hydropiper*
cliver/s/	Arctium, Galium aparine
cockle	Agrostemma githago, Lolium temulentum, Nigella sativa*
lion's foot	Alchemilla, Helleborus niger*
mallow	Althaea, Malva
ray	Agrostemma githago,* Lolium temulentum
sanguinary	Achillea millefolium, Capsella bursa-pastoris, Digitaria sanguinalis*, Polygonum hydropiper
scaryole	Cichorium endivia, Lactuca serriola (?)
setwall	Curcuma zedoaria, Trigonella foenum-graecum*
stanmarch	Apium graveolens,* Smyrnium olusatrum
wallwort	Parietaria diffusa,* Sambucus ebulus
wild garlick	Allium ursinum, A. vineale
woodbind	Convolvulus, Lonicera

---

[28] As given sub "De terbentine". *Pinus* is usually *pyne tree* in GH.

5. *The OED and the English plant names in the Herball*

As is well known, the OED is uneven in period coverage. On the whole, the early 16th century belongs to the periods "currently under-represented in the O.E.D. documentation" (Schäfer 1980, p. 8). *The Grete Herball* (1526) yields, in fact, a great number of plant names (83) antedating the earliest OED entries.[1] Half of these names have however been recorded elsewhere as Middle English (see MED; also Brodin 1950) or, occasionally, as Old English (see MED and Bierbaumer 1975–79) or were noted as GH names by Britten & Holland (B–H). One name (*garden mint*) is first found in Banckes's herbal (1525). Such names will be listed separately after my own GH antedatings to the OED, which are:

agriot (OED 1611)
bear's foot (1551; also in Turner 1538)
bistort (1578)
black bryony (1805)[2]
camel's straw (1578)
case-weed (1578)
citron (1530) (cf. note in Overall list)
damask plum (1616)
dog rose (1597 as 'Rosa canina')
gall nut (1572)
garden ginger (1597)
goosefoot (1548)
great bur (1562)
hare's ballocks (1562)
hare's lettuce (1607)
hog's meat (1756)
holy thistle (1587)
king's crown (1597)
knotgrass (1538)
lentils of the water (1548)
lingwort (1538)
little bur (1585)
meu (1548)
oleandre (1548)
parsley of Macedonia (1578)
polytrich (1725)
pompion (1545)
quince apple (1600)
sea onion (1548)

---

[1] None of them is in Bailey 1978 and the OED supplements "exclude, in the main, pre-1820 antedatings of *O.E.D.* words or senses from general English sources" (First Supplement to OED (1972), p. xv). For some 60 antedatings and additions to OED from *The vertuose boke of Distyllacyon* (1527), see Rydén forthcoming.
[2] Noted by B–H as occurring in Lyte 1578.

snakegrass (1883)
spewing nut (1586)
spike Celtic (1540)
sugar reed (1719)
swine bread (1591)
water flag (1578)
white bryony (1832)[3]
white vine (1542)
wild leek (1551)
wild rape (1551)
wild smallage (1785)
wild valerian (1548)
yellow flag (1550)

The following plant names are GH antedatings to the OED, antedated elsewhere (cf. above) or are names noted as GH names in Britten & Holland (B–H):

aron (OED 1611)
agaric (1533)
alkanet (1567)
alleluia (1543)
apple of Paradise (1676)
arsmart (1551)
bitter almond (1622)
black ivy (1578)
bugloss (1533)
ceterach (1551)
churl's treacle (no OED date)
dove's foot (1548)
dropwort (1538)
duck('s) meat (1538)
elf-dock (1879; noted as GH name in B–H)
eupatory (1542)
five-leaved grass (1578; noted as GH name in B–H)
garden mint (1530; Banckes's herbal 1525)
goat's leaf (1861)
goosebill (1597; noted as GH name in B–H)
hare's beard (1597; noted as GH name in B–H)
hare's palace (1607; noted as GH name in B–H)
hig(h) taper (1548; noted as GH name in B–H)
Indian nut (1613)
knee holm(e) (1562)
knotwort (1845)
liquorice tree (1548)
mare's tail (1762)
nenuphar (1533)
oak fern (1548)

---

[3] Noted by B–H as occurring in Lyte 1578.

paigle (1530; noted as GH name in B–H; cf. Overall list)
red madder (1597)
Solomon's seal (1543; noted as GH name in B–H)
sorb (1530)
spike (1539)
spinach (1530)
squinant (1548)
white poppy (1578)
wild cress (1562)
wild hemp (1597)
wild mint (1578)

As remarked by Schäfer 1980 (p. 66), "antedatings of five, ten, twenty or even thirty years will scarcely influence our views on the history of the English language", whereas "antedatings of half a century or more ... begin to affect our concepts of periods". If we consider the chronological spread of my 42 'genuine' GH antedatings of OED first citations, we discover that they have a range of 4 to 357 years (1530–1883), though the great majority (some 76%) are antedatings of 12 to 70 years only (see below). Within these years (1538–1600) the important botanical works by Turner, Lyte and Gerard were published, authors with a high innovation rate for plant names. The point here is that many plant names "first attested" in the great herbalists, and in some cases considered to be coined by them (cf. p. 40), can be shown to have a longer tradition in the language.[4] And for Turner, in particular, the *Herball* probably served as a direct plant-name source to a greater extent than has been assumed.[5]

Chronological spread of GH antedatings to the OED (not antedated elsewhere):

1530–1550	13 names
1551–1600	19 names
1601–1650	3 names (agriot, damask plum, hare's lettuce)
1651–1700	—
1701–1750	2 names (polytrich, sugar reed)
1751–1800	2 names (hog's meat, wild smallage)
1801–	3 names (black bryony, white bryony, snakegrass)

The inclusion of the supplementary list of antedatings (*aron-wild mint*) does not change the distributional picture materially.

In addition to the antedatings accounted for above, there are in *The Grete Herball* (1526) a number of plant names not recorded in the OED, though a few of them are noted in Britten & Holland (B–H). However, most of these

---

[4] Of the 83 GH antedatings to OED, 39 were previously first attested in Turner (1538, 1548, 1551–68), Lyte (1578) or Gerard (1597).
[5] Otherwise Turner seems to have been very sceptical of the *Herball* (see Hoeniger & Hoeniger 1969a, p. 23; cf. also above, p. 46, and Green 1914, pp. 20f.).

names are mere translations, adaptations or adoptions of plant names as appearing in *Le grant herbier* (cf. above 2 C), very few of which have lived on into modern English (see Chapter 6). GH names not found in the OED are:[6]

affodylly (word form only)[7]
artetyke (noted as GH name in B–H)
bear's bough/twig
blue flower-de-luce
bright (noted by B–H as occurring in GH and in Gerard 1597)
chervell 'Lonicera' (noted as GH name in B–H)
chibol of the sea (chibol is ME)
coronary
cowgourde
crayfery (noted as GH name in B–H)
croyt marine
date of India
earth-thought
fume of the earth (noted as GH name in B–H)
gangelon
garden mallow
grass of the field
Alexander's gourd
great centaury (noted by B–H as occurring in "black-letter herbals")
great mugwort
hare trefle
he-fern (cf. OED male fern 1562)
herb of India
herb of musk
herb Philip
Hercules' grass
high mallow
less bur
less skirret
little clote
little plantain
Margetym gentyll (the form)
meek galingale[8]
middle mugwort
more morel (cf. Brodin 1950, p. 305)
nespyte (noted by B–H as occurring in Gerard 1597)
orient saffron
pyllulary
remcope
small mugwort
smoke of the earth (noted as GH name in B–H)
strofulary

[6] For plants designated as "common" or "tame", see pp. 23 and 97.
[7] See the Overall list.
[8] Also in Banckes's herbal (1525) and in Gerard 1597 (in "A Supplement or Appendix").

tamaryte (the form)
three-cornered rush
tintymall (the form)
tintymall of Babylon
water onion
wild aloe
wild apple
wild blackberries
wild borage (cf. B–H, p. 514, and Brodin 1950, p. 305)
wild galingale
wild mulberries
woodyp

GH *oker* 'acorn' (sub "Arbor glandis latine") is not in the OED. Nor is *waterwort* (GH 1561) for *Crithmum maritimum.*

6. The current English plant nomenclature and the English plant names in the Herball

One of the objects of plant-name research is to analyse the change (discontinuity) or non-change (continuity) in plant-name use and in plant-name equivalence ("synonymy") through time. In the present study, plant-name usage in *The Grete Herball* (1526) has been compared with the English plant nomenclature as given in a modern standard flora, viz. Clapham–Tutin–Warburg 1962, which can be said to represent, in the main, the current, "officially" accepted, nomenclature of wild or naturalized British plants. This flora was complemented with a more popular (and widely used) flora of vascular plants, viz. McClintock & Fitter 1955 (1971), which however added very few names to those included in Clapham–Tutin–Warburg.[1]

Of the some 500 English names for plants (in a wide sense) in the *Herball* about 175 or 35 % are represented, as simplexes or as part of compounds or phrasal names, in Clapham–Tutin–Warburg 1962. These names are given below. The Latin names adduced indicate PE botanical reference. Names with botanical references differing from those in GH are asterisked.

agrimony (Agrimonia)
alexanders (Smyrnium olusatrum)
alkanet (Pentaglottis sempervirens)*
almond (Prunus Amygdalus)
apple (in crab apple, Malus sylvestris)
archangel (yellow) (Galeobdolon luteum)*

[1] A collation with the normative lists given in Dony *et al.* 1974 did not yield any additional names.

ash (Fraxinus excelsior)
baldmoney (Meum athamanticum)*
balm (Melissa officinalis)
barberry (Berberis vulgaris)
barley (Hordeum/Hordelymus)
bear's foot (Helleborus)* (?)
beet (Beta vulgaris)
betony (Betonica officinalis)
bistort (Polygonum bistorta)
blackberry (Rubus)
borage (Borago officinalis)
box (Buxus sempervirens)
bramble (Rubus)
bright (in eyebright, Euphrasia)*
broom (Sarothamnus scoparius)
bryony (Bryonia)
buck's horn (plantain) (Plantago coronopus)*
bugloss (Anchusa arvensis) Viper's bugloss is Echium vulgare
bullace (Prunus domestica, ssp. insititia)
calamint (Calamintha)
celandine (Chelidonium majus/Ranunculus ficaria)
centaury (Centaurium)
cherry (Prunus)
chervil (Anthriscus cerefolium)
chestnut (Castanea)
chicory (Cichorium intybus)
cinquefoil (Potentilla)
clary (meadow, wild) (Salvia)
cleavers (Galium aparine)
cole (Brassica)
comfrey (Symphytum)
coriander (Coriandrum sativum)
(corn) cockle (Agrostemma githago)
cowslip (Primula veris)
cress (Rorippa, Cardamine, etc.)
crowfoot (Ranunculus)
cuckoo-pint (Arum maculatum)
currant (Ribes)
cypress (Cupressus)
[daffodil]
daisy (Bellis perennis)
darnel (Lolium temulentum)
dead nettle (Lamium)
devil's bit (scabious) (Succisa pratensis)
dock (Rumex)
dodder (Cuscuta)
dropwort (Filipendula vulgaris)
duck's meat (Lemna minor)
dwale (Atropa belladonna)
elder (Sambucus)

felwort (Gentianella amarella)
fennel (Foeniculum vulgare)
fenugreek (Trigonella)
fern (Polypodiaceae)
fig (Ficus)
filbert (Corylus maxima)
fir (Abies)?
flax (Linum usitatissimum)
fumitory (Fumaria)
galingale (Cyperus longus)*
garden cress (Lepidium sativum)
garlic (Allium sativum)
gentian (Gentianella)
germander (Teucrium)
gladdon (Iris foetidissima)* (?)
goosefoot (Chenopodium)*
great plantain (Plantago major)
gromwell (Lithospermum)
groundsel (Senecio vulgaris)
hart's tongue (fern) (Phyllitis scolopendrium)
hemlock (Conium maculatum)
hemp (Cannabis sativa)
henbane (Hyoscyamus niger)
hollyhock (Althaea rosea)
honeysuckle (Lonicera periclymenum)
hop (Humulus lupulus)
horehound (white) (Marrubium vulgare)
horsemint (Mentha longifolia)
hound's tongue (Cynoglossum)
houseleek (Sempervivum tectorum)
hyssop (Hyssopus officinalis)
ivy (Hedera helix)
juniper (Juniperus communis)
knotgrass (Polygonum aviculare)
leek (Allium)
lettuce (Lactuca)
lily (Lilium)
lovage (Ligusticum scoticum)*
lungwort (Pulmonaria)
madder (Rubia tinctorum)
maidenhair (Adiantum)
mallow (Malva)
mare's tail (Hippuris vulgaris)
[marigold (Calendula)]
mayweed (Anthemis/Matricaria/Tripleurospermum)
medlar (Mespilus germanica)
melilot (Melilotus)
mercury (Chenopodium bonus-henricus/Mercurialis)
meu (Meum athamanticum)
milfoil (Achillea millefolium)

millet (Setaria italica)*
mint (Mentha)
motherwort (Leonurus cardiaca)*
mouse-ear (Cerastium/Hieracium)
mugwort (Artemisia/Galium cruciata)
mullein (Verbascum)
musk (Mimulus moschatus)
nettle (Urtica)
nightshade (Solanum dulcamara/nigrum)
oak (Quercus robur)
oak fern (Thelypteris dryopteris)*
oats (Avena)
onion (Allium cepa)
orache (Atriplex)
oxtongue (Picris)*
paigle (Primula veris/elatior)
parsley (Petroselinum)
pear (Pyrus communis)
pellitory (Parietaria)*
pennywort (Hydrocotyle vulgaris/Umbiculus rupestris)*
peony (Paeonia)
periwinkle (Vinca)
pimpernel (Anagallis/Lysimachia nemorum)*
pine (Pinus)
plantain (Plantago)
plum (Prunus domestica)
polypody (Polypodium vulgare)
poppy (Papaver)
purslane (Portulaca oleracea/Halimione (sea purslane))
quince (Cydonia oblonga)
radish (Raphanus)
ramsons (Allium ursinum)
rape (Brassica napus)
ray (-grass) (Lolium)
ribwort (Plantago lanceolata)
rose (Rosa)
rue (Ruta)
rush (Juncus)
rye (Secale cereale)
savory (Satureja montana)*
saxifrage (Saxifraga/Pimpinella saxifraga (burnet saxifrage))*
scabious (Knautia/Succisa)
self-heal (Prunella vulgaris)*
shepherd's purse (Capsella bursa-pastoris)
sloe (Prunus spinosa)
Solomon's seal (Polygonatum multiflorum)
sorrel (Rumex)
sowthistle (Sonchus)* (?)
spurge (Euphorbia)
squill (Scilla)

St. John's wort (Hypericum)
sti(t)chwort (Stellaria)
strawberry (Fragaria)
teasel (Dipsacus)
thistle (Carduus, Cirsium, etc.)
tormentil (Potentilla anglica/erecta)*
trefoil (Trifolium)
tutson (Hypericum androsaemum)*
valerian (Valeriana)
vervain (Verbena officinalis)
violet (Viola)
wallnut (Juglans regia)
watercress (Rorippa)
wheat (Triticum)
wild leek (Allium ampeloprasum)*
willow (Salix)
woodruff (sweet) (Galium odoratum)* (?)
woodsorrel (Oxalis acetosella)
wormwood (Artemisia)
yarrow (Achillea millefolium)
yellow flag (Iris pseudacorus)*
yew (Taxus baccata)

In addition, there are some 35 GH names which are standard today but not found in Clapham–Tutin–Warburg, including words such as *mushroom, toadstool* and *moss*, some names for cultivated plants like *dill, myrtle* and *rhubarb* and some words for fruits and spices like *canel, caper, clove* and *ginger*. This gives us a total of some 210 GH words for plants, fruits and spices (or ca. 40 % of such words as mentioned in GH) which are in "standard" use today, though not necessarily with the same botanical reference as that given in the *Herball.*

A noteworthy feature is that the overwhelming majority of the GH names that are part of the current standard nomenclature are of Old or Middle English descent. Of the 124 names first or only attested in the *Herball* (cf. above 2 C) only the following are in Clapham–Tutin–Warburg (names with *common* disregarded): *bear's foot, bistort, bright* (in *eyebright*), *goosefoot, great plantain, knotgrass, meu, Solomon's seal, wild leek, woodsorrel* and *yellow flag.* Few of the numerous translations, adaptations or adoptions of French or Latin names in the *Herball* or of the GH names not attested in the OED have caught on in the language, at least in the standard nomenclature.[2]

Not a few names in the *Herball*, though not part of the "official" nomenclature of today as mirrored in Clapham–Tutin–Warburg and McClintock & Fitter, have been noted as in local or regional use. A comparison with Grigson 1955 (1975), which to my knowledge offers the most extensive printed lists of

[2] For these names, see above 2 C (translations, etc.) and 5 (names not recorded in OED).

English dialect plant names[3]—Grigson's "Index of local names" comprises some 5 200 items—yields an additional 52 names to those included in the standard floras consulted.[4] This totals the number of GH words for plants (incl. fruits/spices) that have lived on (in one sense or other) into modern times, as "standard" and/or "local", to some 260 or about half of the "English" plant names in the *Herball*. It should be emphasized that the temporal range of the names included in Grigson's lists is wide (some 100 years) and that, consequently, the inclusion of a name in his lists "does not guarantee that it is still in use" (Grigson 1975, p. 25). Grigson's chief sources were the OED, Britten & Holland, Wright's dialect dictionary and county floras.

As stated above, the great majority of the GH names now used as standard names are of Old or Middle English heritage. This also applies to the 52 "local" names, of which only the following are GH first records, i.e. not attested in OE/ME sources or in 16th century records prior to the *Herball*:

carpenter's grass, case-weed, cuckoo's bread, cuckoo's meat, frog's foot, goosebill, hare's beard, hare's lettuce, hig(h) taper, St. Peter's wort, sea onion, snakegrass and swine bread. For plant references of these names in dialects, see Grigson 1975.

To sum up: of the some 260 GH names in standard or local use today (for currency of the local names, cf. above) only 25 represent names as first recorded in the *Herball*, whereas of the some 375 Old or Middle English names in the *Herball* about 195 or 52 % have lived on into our own time, as standard or local.

The value of the figures obtained here for plant-name longevity in English is difficult to assess in the absence of any parallel investigations.[5] It is obvious, however, that the names current or at least with some tradition in the language in the early 16th century have, on the whole, proved far more successful than those invented, translated or adopted by the writer of the *Herball*.[6] This does

---

[3] The "Index" of synonymous plant names submitted in Britten & Holland (pp. 563–618) is panchronic and supplies lists containing book-names as well as folknames.

[4] These names are: ache, affodil, alleluia, anise, arsmart, blind nettle, bloodwort, bonewort, bruisewort, calf's foot, caltrop, carpenter's grass, case-weed, clote, cockle (in Clapham–Tutin–Warburg only in cockle bur and corn cockle), cuckoo's bread, cuckoo's meat, earthgall, eglantine, five-leaved grass, frog's foot, goat's leaf, goosebill, hare's beard, hare's lettuce, hig(h) taper, hindheal, knee holm, line, lion's foot, maithen, open arse, organ, St. Peter's wort, sea onion, sengreen (and variants), snakegrass, sparrow tongue, spikenard, swinebread, swine's grass, wallwort, wartwort, waybre(a)d, woodbine, wild borage / mint / parsley / rue / sage / teasel / thyme.

[5] On the continuation of "Linnaean" names in modern Swedish floras and dialects, see Fries 1962 and 1977 and Vide 1962.

[6] Of plant names first recorded in GH which have ousted (in standard PE) their earlier attested equivalents may here be mentioned *knotgrass* (for *Polygonum aviculare*), *woodsorrel* (for *Oxalis acetosella*) and *yellow flag* (for *Iris*).

not mean, of course, that early translations or modifications of foreign names have not, as a whole, survived into the current language—many names, like *cornflower, loosestrife, twayblade* and *wintergreen*, coined by Turner, Lyte, Gerard (and others) are evidence to the contrary. It only implies that such names as attempted in the *Herball* have usually proved too un-English or have simply not been brought into fashion by influential people (botanists or others) or books. And, self-evidently, the existence of a GH name in later English does not necessarily imply continuation in usage, in terms of direct GH lineage. A case in point here, among others, is *snakegrass* which is evidenced in the *Herball* as a mere translation (referring to *Dracunculus vulgaris*), but which is also used, with reference to other plants, in modern dialects (cf. Grigson 1975, pp. 98 and 306).

Plant Names in The Grete Herball (1526) as Part of Chapter Headings

The sum total of English plant names (names of fruits, etc.) as part of chapter headings in the *Herball* is 315,[1] or 63 % of the English plant names in the book, 41 of which are GH first or unique records (asterisked). 49 (or 18 %) of the headings include 2 variant English names. Two of the headings ("De Enula campana" and "Zizania") contain 3 names, one ("De lingua passerina") 4. The order of the items as given in the *Herball* is retained here.

De Agno casto	Tutson
De Apio	smalache or stammarche
De Apio ramio	wylde smalache*
De Apio risus	Crowfote or ache
De Achasio	iuce of Sloes or bolays
De Aneto	Dyll
De Affodillio	Affodylly*
Allium latine. Scordon vel scordeon grece. Thaū Arabice	Gorlyke [for Garlyke]
Acorus	Gladon
Anisum latine & grece Aneisum Arabice	Anys
Absinthium latine. Grece absinthion. Saxicon Arabice	Wormwood
Amigdala latine. Lanet Arabice & grece	Almondes
De Aristologia rotunda …	Smerewort or meke galyngale
Aristologia longa latine	Reed mader
De Arthemesia	Mugwort or moderwort
De Arthemesia minor	Of the myddle mugwort*
De Arthemesia minima	the lesse mugwort*
De alcamia	Alcamet
Adianthos	Maydenwede
De agrimonia	Egrymony
De Altea	Malowe
De astula regia	Woodroue
De Ambrosiana	Hyndhele
De Atriplice	Arache
De Auena	Otes

[1] The following names appear more than once: balm (tree) (2), beet (2), chicory (2), cliver (2), cockle (3), dill (2), long plantain (2), mallow (2), pedelyon (2), setwall (2), skirret (2), wallwort (2), wild lettuce (2), woodbind (3), yarrow (2).

Ameos	Woodnep*/or peny wort
De Alleluya	Wood sorell* or cukowes meate*
Acetosum latine. Numa Arabice. Oxiolapatium Grece	Sorell
Auelana	Fylberdes
De Albatra	Tormentyll
De Balsamo	Bawme tre
De balaustia	floures of pomgarnatis
De Boragine	Borage
De baucia	Skyrwyt
De Bethonica	Bethony
De lingua anseris	Goosbyll*/or styche wort
Brancha vrsina	Bearefote*
De Berberis	Berberies
De buglossa	Oxtongue /or langdebefe
De Berbena	Uernayne
De bursa pastoris	Cassewed*
De brionia	Wylde neppe or bryony
De bedegar	Eglentyne
De bardana	A clote that bereth burres
De buxo	Box tre
De bleta	Betes
Capilli veneris	Maydī here
De Cipresso	Cypresse
De cinamomo	Cynamome/or canell
De camedrios	Germaundre
De camephitheos	Mederacle*
De Cimino	Comyn. sede
De cicuta	Hemlocke
De Croco	Saffron
De Cepe	Onyon
De Costo	Cost mary
De Cucurbyta	A gourde
De Cucumero	Coucōmers [cucumbers]
De Cyt[r]ullo	Cytrons* or cytrulles
De Celydonia	Celendyne
Colloquintida	wilde gowrde
De cuscuta	Dodyr
De calamento	Calamynt
De centaurea	Centory
De coryandro	Coryandre
De caules	Caule wortes
De cerifolio	Cheruell
De Canapis	Hempe
De cameleonta	Wolfe thystle
De camomylla	camomylle
De Castaneis	chestnuttes
De catapucia	Spourge
De canna	A rede
Calendula	Mary gowles/or ruddes

66

De Consolida maiori	Comfrey
Consolida media	Maythen
De consolida minori	Dyasy [for Daysy] or brusewort
Coronaria	Honysocle
De cerasis	Cheryes
De caprifolio	Woodbynde
De Dauco	Dawke
De Diptano	Dytany
De Dactilis	Dates
De endiuia	Endyue
De Enula campana	Elfe docke:* Scabwort or horshele
De eupatorio	Wylde sawge
De epatica	Lyuerwort
De eleboro albo	Lyngwort*/or peleter of Spayne
De elleboro nigro	Pedelyon/or lyons fote
De Eruca	Skyrwyt. Or wylde cawles that bered [for bereth] mustarde sede
De Ebulo	Walworde
De edera magna	yuy
De spatula fetida	yelowe flagge*
De Elitropio	Chycory
De eufragia	eufrace
De flamula	sereworde [for spereworde]
De fumo terre	Fumyterry
De Filipendula	Dropwort
De Fraxino	Asshe tre
Feniculus latine. Hazienis vel Hakasmech Arabice	Fenell
De Fenegreco	Fenegreke or setwall
De filice	Ferne
De Fragaria	S[t]rawberyes
De faba communi	Beanes
De fungis	Mussherons
De filice dicto os munda	Heferne*
De Fycu	Fygges
De gariofilis	Clowes
De genciane	Felwort or baldymony
De galanga	Galyngale
De Gariofilata	Auens
De herba Indica	Gith. Cokyll
De Milio solis	Gromyll/or lychwale
De gallitrico	Clarey
De galla	Galles nuttes (the fruite of okes)
De genestula	woodyp*
De genesta	Brome
De gramine	Quekes
De herucaria	Wartwort
De herba paralisi	Cowslyp or pagle
De Iusquiamo	Henbane
De ysopo	ysope

De Iaro	Cuckowe pyntyll
De Ire	Bleweflourdelyce*
De Ipoquistidis	Tode stoles
De Iunipero	Ienepre
De Iperyco	Herbe Johñ/or saynt Johannis worte
De Lambrusca	Wylde wyne
De lilio	Lylly
De lingua auis	Asshe sede
De mercuriali	Mercury
De Lapacio	Reed docke
De Lactuca	Letuse
De Lactuca siluestri	Wylde letuse
De lauro	Laurel or bayes
De laureola	Rybbwort
De leuistico	Louage
De lolio	Cokyll
De lupulo	Hoppes
De pede leonis	Pedelion
De lactuca agresti	Wylde letuse
De femine lini	Lyne sede
De lentycula aque	Grenes*/or ducke meate
De cynoglossa	Hondestonge
De lingua hircina	Buckesshorne
Lanceolata	Longe plantayn*
De lactuca leporina	Hares letuse*
De lapaceola	Lytell burre* or clyuer
De malua	Malowes
De maluauisco	Wylde malowes
De malua ortulana	Holyhocke
De menta	myntes
De menta romana	Wytmynt
De mentastro	Horsmynt
De Mandragora	Mandrake
De citoniis	Quynces
De granatis	Pomgarnades
De macianis pomis	Wood crabbes or wyldynges*
De Marrubium	Horehounde
De Mace	Maces
De Milio	Mylle
De maiorana	Margetym gentyll*
De melissa	Bawme
De mora celsi	Molberyes
De matrisilua	Wood bynde
De Petrocilio macedonico	Stammarche or Alysamder
Morsus diaboli	Remcope* or deuylles bytte
De millefolio	yarowe/myllefoyle
De Melonibus	Melons
De Narsturcio	Tame cresses
De Narsturcio agresti	Wylde cresses
De nuce muscate	The Nutmygge

De nuce Indica	Nuttes of Inde
De nuce communi	Wall nuttes
De nuce vomyca	Spewynge nuttes*
De Nigella	Cokyll
De Nespilis	Medlers or open arses
De Origano	Brotherworte
De Ordeo	Barly
De Oliua	Olyues
De piretro	Walworte
De pipere	Peper
De peonia	Pyony
De Papauere	Poppy
De peucedano	Dog fenell. The mydde consolde[2]
De petrocilio	Percely
De pineis	Pyne trees or apples
De prunis	Plommes
De polipodio	Oke ferne
De portulace	Porcelayne
De polio montano	Wylde tyme
De plantagine	Plantayne or weybrede
De lanceolata	Longe plantayne (*)
De panico	Panyke
De Penthafilone	Synkefoyle/or .v. leued grasse
De lingua passerina	Sentynode. swynes grasse knot-grasse*/or sparow tongue
De polytryco	Walfarne*
De primula veris	Prymerolles
De palacio leporis	Hares palays*
De pulmonaria	Crayfery* or lungwort
De Persicaria	Arssmert or culrage
De pimpinella	Selfe heale or pympernell
De pilocella	Mows eare
De prouinca	Perwynke
De pede columbino	Doues fote
De ruta	Rue
De Rosa	Rose
De raffana	Rape rote
De radice	A radysshe
De Reubarbaro	Rewbarbe
De Rubea	Madder
De Porro	A leke
De piganio	Wylde rue
De rose marino	Rosmary
De rubo	a brere or bramble
De Riso	Rys
De Rapiastro	Wylde rapes*
De rapa	Rapes

[2] Equivalent of *dogfennel* = *Anthemis cotula*. Not in the appended text. There is no parallel name in the French text.

De spicnardo	Spyknarde or spyke
De solatro	Petymorel. or nyght shade
De solatro rustice	Dwale or more morell
De semper viua	Howsleke or selfegrene
De Saponarya	Crowsoppe
Squinanto	Camelles strawe*
De semine napei	Musterde sede
De satirione	Gangelon* or hare ballockes*
De Cichorea	Chycory
Scordeon	Wylde garlyke
De sperago	Sperage
De Sauina	Sauyn
De Saxifraga	Saxifrage
De Saluia	Sawge
De scabiosa	Scabyous
De narsturcio	Cresses
De senacionibus	Grownswell
De serpentina	Dragons/or snakegrasse*
De salicibus	A wyloue tree
De sambuco	Eldre
De squilla	A squyll or see onyon*
De serpyllo	Pellyter
De satureia	Sauerey
De sanguina[r]ia	Blodworte/or yarow
De stolopendria	Hertes tongue
De spinachia	spynache
De sicla/alias bleta	Betes
De stalogia	Cyues
De spergula	Clyuers
De silfu	wylde valeryan*
De sistra	Dyll
De Salunica	Caltrappe
De Sigillo sancte marie	o[u]r ladyes saele
De Saxifraga minori	The lesse saxifrage*
De Sinomo	Wylde percely
De tapso barbato	Hareberde*/or hygtaper*
De tribulo marino	Reed brere
De trifolio	Trefle or thre leued grasse
De frumento	wheate
De violis	Vyolettes
De valeriana	Valerian
De virga pastoris	Wylde tasyll
De viperina/alias vrtica mortua	Deed nettel or archaungell
De vrtica	Nettle
De Uolubilis	Woodbynde
De Uua	A grape
De vngula caballyna	lytell clote*
De zinzibre	Gynger
De Zedoare	Setwale
Zizania	Ray/drawke/darnell

Arbor glandis latine. Hullus Arabice	An oken tree
De Siligo	Rye
Pira	Peres
Poma	Apples
Vsnea. vel muscus arborum	Mosse
De Cardone benedicto	Sowthistle
Citrum	A tre so named
Uua passe	Rasyns of carans [for corans]
Vibex	a byrten tre[3]

[3] Cf. OED/MED birthel/tre/ 'fruit-bearing tree'. The GH record post-dates the OED/MED citations.

The English Plant Names in The Grete Herball 1526 (Overall list)

Items commented on in the Notes are starred. GH form/spelling is (as a rule) as given in headings or, when a name is not part of a heading, as first attested in the text. Names of fruits and berries are usually in the plural. Abbreviations (e.g. ē for *en* or m̄ for *mm*) have been silently expanded. "ME" denotes that the word (irrespective of bot. ref.) is first recorded in Middle English (c. 1150–1500), "OE" that the word (irrespective of bot. ref.) is recorded in Old English. Items not marked "OE" or "ME" are, unless otherwise stated, GH first records (cf. Chapters 2C and 5). The "modern" form given does not necessarily imply current use of the name. And that a GH name is paralleled with a "Latin heading" does not necessarily imply that it is part of the heading, only that it occurs sub that heading (as part of the heading and/or the concomitant text). Identification (plant and/or fruit, seed, etc.)—in some cases only tentative—is in terms of the modern scientific name (basically according to Clapham–Tutin–Warburg 1962 and Turner 1965).

Name in GH	Modern form	Latin heading in GH	Modern scientific name
aaron (ME)*	aron	De Iaro	Arum maculatum
ache (ME)*	ache	De Apio risus	Ranunculus sceleratus
affodylly*	daffodil	De Affodillio	Asphodelus
agaryk (ME)	agaric	De Agarico	Polyporus officinalis
agryotes*	agriot	De cerasis	Prunus (a kind of cherry)
alcamet (ME)*	alkanet	De alcamia	Alkanna tinctoria/ Lawsonia inermis
alleluya (ME)*	alleluia	De Alleluya	Oxalis acetosella
almondes (ME)	almond	Amigdala latine	Prunus Amygdalus
aloe(n) (OE)*	aloe	De Aloe/De ligno Aloes	Aloe
alysamder (OE)*	alexanders	De Petrocilio macedonico	Smyrnium olusatrum
anet (ME)	anet	De Aneto	Anethum graveolens
anys (ME)*	anise	Anisum latine & grece	Pimpinella anisum
apples (OE)	apple	Poma	Malus
apple of paradys (ME)	apple of Paradise	De Musis	Musa paradisiaca
arache (ME)	orach(e)	De Atriplice	Atriplex hortensis
archaungell (ME)	archangel	De viperina/alias vrtica mortua	Lamium

Name in GH	Modern form	Latin heading in GH	Modern scientific name
arssmert (ME)	ars(es)mart	De Persicaria	Polygonum hydropiper
artetyke*		De herba paralisi	Primula veris
asshe tre (OE)	ash	De Fraxino	Fraxinus excelsior
auelane (ME)	avellan	Auelana	Corylus maxima
auens (ME)	avens	De Gariofilata	Geum urbanum
baldymony (ME)*	baldmoney	De genciane	Gentianella
barley (OE)*	barley	De Ordeo	Hordeum vulgare
basyll (ME)	basil	De basilicone	Ocimum basilicum
bawme (tre) (ME)	balm	De Balsamo/De melissa	Commiphora opobalsamum/ Melissa officinalis
bayes (ME)	bay	De lauro	Laurus nobilis
beanes (OE)	bean	De faba communi	Vicia faba
beares bough / twygge	bear's bough / twig	Brancha vrsina	Acanthus mollis
bearefote*	bear's foot	Brancha vrsina	Acanthus mollis
berberies (ME)*	barberry	De Berberis	Berberis vulgaris
betes (OE)	beet	De bleta / De sicla alias bleta	Beta vulgaris
bethony (ME)*	betony	De Bethonica	Betonica officinalis
bistorte	bistort	De bystorta	Polygonum bistorta
blacke beryes (OE)	blackberry	De Mora bacci	Rubus fruticosus (the fruit)
blacke bryony*	black bryony	De brionia	Tamus communis
blacke elebore (ME)	black hellebore	De elleboro nigro	Helleborus niger
blacke yuy (ME)*	black ivy	De edera magna	Hedera helix
blacke poppy (ME)	black poppy	De Papauere	Papaver somniferum
blewe flourdelyce*	blue flower-de-luce	De Ire	Iris germanica (?)
blodworte (ME)*	bloodwort	De Persicaria / De sanguina[r]ia	Polygonum hydropiper / Achillea millefolium
bolays (ME)	bullace	De Achasio	Prunus domestica, ssp. insititia
bonewort (OE)*	bonewort	[Adianthos]	Bellis perennis (?)
borage (ME)*	borage	De Boragine	Borago officinalis
box (tre) (OE)	box	De buxo	Buxus sempervirens
bramble (OE)	bramble	De rubo	Rubus fruticosus
brere (OE)*	briar	De rubo	Rosa
brome (OE)	broom	De genesta	Sarothamnus scoparius
brotherworte (OE)*	brotherwort	De Origano	Origanum vulgare (?)
brusewort (OE)	bruisewort	De consolida minori	Bellis perennis
bryght*	bright (eye)	De Celydonia	Ranunculus ficaria (?)
bryony (ME)	bryony	De brionia	Bryonia dioica
buckesshorne (ME)*	buck's horn	De lingua hircina	
buglosse (ME)*	bugloss	De buglossa	Anchusa (or "similar" plants)
burit (ME)*	borith	De Saponarya	Saponaria officinalis
b[l]ynde nettell (OE)*	blind nettle	De viperina / alias vrtica mortua	Lamium
byrten tre (ME)*		Vibex	

Name in GH	Modern form	Latin heading in GH	Modern scientific name
bytter almondes (ME)	bitter almond	Amigdala latine: of bytter Almondes	Prunus Amygdalus (v. amara)
calamynt (ME)*	calamint	De calamento	Calamintha
caltrappe (ME)*	caltrop(s)	De Salunica	Centaurea calcitrapa
calues fote (ME)*	calf's foot	De Iaro	Arum maculatum
camelles strawe*	camel's straw	Squinanto	Cymbopogon schoenanthus
camomylle (ME)	c(h)amomille	De camomylla	Chamæmelum nobile
canell (ME)	canel	De cinamomo	Cinnamomum
canne (ME)*	cane	Enula campana	
capparis (ME)*	caper	De Capparis	Capparis spinosa
carpenters grasse	carpenter's grass	De millefolio	Achillea millefolium
carui (ME)	carvy	De Caruo	Carum carvi
cassewed	case-weed	De bursa pastoris	Capsella bursa-pastoris
caule wortes (OE)*	colewort	De Caules	Brassica oleracea
celendyne (ME)*	celandine	De Celydonia	Ranunculus ficaria (?)
cellydony (OE)	celidony	[De eufragia]	Chelidonium majus
centory (ME)	centaury	De centaurea	Centaurium erythræa
ceterach (ME)*		De ceterach	Ceterach officinarum
cheruell*		De caprifolio	Lonicera periclymenum
cheruell (OE)*	chervil	De cerifolio	Anthriscus cerefolium
cheryes (ME)	cherry	De cerasis	Prunus Cerasus
chestayne (ME)	chesteine (chesten)	[De Melle]	Castanea sativa
chestnuttes (ME)	chestnut	De Castaneis	Castanea sativa
churles tryacle* (ME)	churl's treacle	Allium latine	Allium sativum
chybol of the see (see see onyon)			
chyches (ME)*	chich [chickpea]	Cicer	Cicer arietinum
chycory (ME)*	chicory	De Cichorea / De Elitropio	Cichorium intybus
clarey (OE/ME)*	clary	De gallitrico	Salvia sclarea
clote (OE)*	clot(e)	De bardana	Arctium
cloves (ME)	clove(s)	De gariofilis	Eugenia aromatica
clyuer(s) (ME)*	cleavers / cliver(s)	De lapaceola / De spergula	Arctium / Galium aparine
cokyll (OE)*	cockle	De herba Indica / De lolio / De Nigella / De Sizania	Agrostemma githago / Lolium temulentum / Nigella sativa
colloquintide (ME)*		Colloquintida	Citrullus colocynthis
comfrey (ME)* [common + noun]*	comfrey	De Consolida maiori	Symphytum officinale
comyn (ME)*	cumin	De Cimino	Cuminum cyminum
consoulde (OE)* /less/ coronary*	consoud, consound	Consolida Coronaria	
coryandre (ME)	coriander	De coryandro	Coriandrum sativum
cost (OE)	cost	De Costo	Tanacetum balsamita (Balsamita major)
cost mary (ME)	costmary	De Costo	Tanacetum balsamita (Balsamita major)
cotton (ME)	cotton	De bombace	Gossypium

Name in GH	Modern form	Latin heading in GH	Modern scientific name
coucommers (ME)	cucumber	De Cucumero	Cucumis sativus
cowgourde		De Cucumero	Cucumis sativus
cowslyp (OE)*	cowslip	De herba paralisi	Primula veris
crayfery*		De pulmonaria	[a lichen]
cress (OE)*	cress	De narsturcio	Lepidium sativum
crowfote (ME)	crowfoot	De Apio risus	Ranunculus sceleratus
crowsoppe (ME)	crowsoap	De Saponarya	Saponaria officinalis
croyt marine*		De cretano	Crithmum maritimum
cuckowe pyntyll (ME)	cuckoo-pint	De Iaro	Arum maculatum
cuckowes brede	cuckoo('s) bread	De Alleluya	Oxalis acetosella
cukowes meate*	cuckoo('s) meat	De Alleluya	Oxalis acetosella
culrage (ME)	cul(e)rage	De Persicaria	Polygonum hydropiper
cynamone (ME)*	cinnamon	De cinamono	Cinnamomum
cypresse (ME)	cypress	De Cipresso	Cupressus semper-virens
cytrons*	citron	De Cyt[r]ullo	Citrus medica
cytrulles (ME)	citrul	De Cyt[r]ullo / De Cucurbyta	Cucumis citrullus / Cucurbita pepo
cyues (ME)	chives	De stalogia	Allium schoenoprasum
damacenes (ME)*		De cerasis	Prunus (a kind of cherry)
damassons (ME)	damson	De prunis	Prunus domestica
damaske plommes	damask plum	De prunis	Prunus domestica
darnell (ME)*	darnel	Zizania	
dates (ME)	date	De Dactilis	Phoenix dactylifera
date of Ynde*	date of India	De Tamarinde	Tamarindus indica
dawke (ME)		De Dauco	Daucus carota
deed nettel (ME)	dead nettle	De viperina / alias vrtica mortua	Lamium
deuylles bytte (ME)	devil's bit	Morsus diaboli	Succisa pratensis
docke (OE)	dock	De Lapacio	Rumex
dodyr (ME)*	dodder	De cuscuta	Cuscuta
dogfenell (ME)*	dog fennel	De peucedano	Peucedanum officinale
dogges rose*	dog rose	De Ipoquistidis	[galls on roses]
doues fote (ME)*	dove's foot	De pede columbino	Geranium
dragons (ME)	dragons	De serpentina	Dracunculus vulgaris
drawke (ME)*		Zizania	
dropwort (ME)	dropwort	De Filipendula	Filipendula vulgaris
ducke meate (ME)*	duck's meat	De lentycula aque	Lemna minor
dwale (ME)	dwale	De solatro rustice	Atropa belladonna
dyasy [for daysy] (OE)	daisy	De consolida minori	Bellis perennis
dyll (OE)	dill	De Aneto / De sistra	Anethum graveolens
dytany (ME)*	dittany	De Diptano	Amaracus (Origanum) dictamnus
eglentyne (ME)	eglantine	De bedegar	Rosa rubiginosa
egrymony (ME)*	agrimony	De agrimonia	Agrimonia eupatoria
eldre (OE)	elder	De sambuco	Sambucus nigra
elebore (ME)*	hellebore	De eleboro albo	Veratrum album / Helleborus niger

Name in GH	Modern form	Latin heading in GH	Modern scientific name
elfe docke	elf-dock	De Enula campana	Inula helenium
endyue (ME)	endive	De endivia	Cichorium endivia
erthe galle (OE)	earthgall	De centaurea	Centaurium erythræa
erththought*		De polytryco	Asplenium
eufrace (ME)*	euphrasy	De eufragia	Euphrasia
eupatory (ME)*		De Saluia	Teucrium scorodonia
ewe (OE)	yew	[De Sauina]	Taxus baccata
felwort (OE)*	felwort	De genciane	Gentianella
fenegreke (ME)*	fenugreek	De Fenegreco	Trigonella foenum-græcum
fenell (OE)	fennel	Feniculus latine	Foeniculum vulgare
ferne (OE)	fern	De filice	Pteridium aquilinum
flag (ME)*	flag	[De eupatorio]	Iris (?)
flax (OE)	flax	[De cuscuta]	Linum usitatissimum
frogges fote	frog's foot	De lentycula acque	Lemna minor
fullers grasse	fuller's grass	De Saponarya	Saponaria officinalis
fume of the erthe		De fumo terre	Fumaria officinalis
fumyterry (ME)*	fumitory	De fumo terre	Fumaria officinalis
fygges (ME)	fig	De Fycu	Ficus
fylberdes (ME)	filbert	Auelana	Corylus maxima
fyrre (ME)*	fir	De terbentine	
fyue (.v.) leued /grasse/ (OE/ME)*	five-leaved grass	De Penthafilone	Potentilla reptans
galles nuttes*	gall nut	De galla	[oak-gall]
galyngale (ME)	galingale	De galanga	Alpinia officinarum
gangelon*		De satirione	[some orchid]
gardyne cresse (ME)	garden cress	De Narsturcio	Lepidium sativum
gardyn gynger*	garden ginger	De Diptano	Amaracus (Origanum) dictamnus
gardyn malowes*	garden mallow	De malua ortulana	Althæa rosea (?)
gardyn mynte*	garden mint	De menta	Mentha spicata
gardyn saffron (ME)*		De croco	Crocus sativus
garlyke (OE)*	garlic	Allium latine	Allium sativum
gencyan (ME)*	gentian	De genciane	Gentiana
germaundre (ME)*	germander	De camedrios	Teucrium chamædrys
gith (ME)	gith	De herba indica	Nigella sativa
gladon (OE)	gladdon	Acorus	Iris pseudacorus
goosbyll*	goosebill	De lingua anseris	Stellaria holostea (?)
goosfote*	goosefoot	De sanguina[r]ia	Digitaria sanguinalis
gotes leues (ME)	goats' leaves (goat's leaf)	De caprifolio	Lonicera periclymenum
gourde (ME)	gourd	De Cucurbyta	Cucurbita pepo
gowrde of Alexandry	Alexander's gourd	Colloquintida	Citrullus colocynthis
grasse of the felde	field grass	De gramine	[species belonging to Gramineæ]
grenes*	greens	De lentycula aque	Lemna minor
grete burre	great bur(r)	De lapaceola	Arctium
grete centory*	great centaury	De centaurea	Centaurea (?)
grete mugwort	great mugwort	De Arthemesia	Artemisia

Name in GH	Modern form	Latin heading in GH	Modern scientific name
grete plantayne*	great plantain	De plantagine	Plantago major
gromyll (ME)*	gromwell	De Milio solis	Lithospermum officinale
grownswell (OE)*	groundsel	De senacionibus	Senecio vulgaris
gylofre (ME)*	[gilly flower]	De cinamomo / De Gariofilata	Dianthus caryophyllus (?)/ Geum urbanum
gynger (ME)*	ginger	De zinzibre	Zingiber officinale
hare ballockes*	hare's ballocks	De satirione	[some orchid]
hare berde*	hare's beard	De tapso barbato	Verbascum thapsus
hare trefle*	hare's trefoil	De trifolio	Trifolium
hares letuse*	hare's lettuce	De lactuca leporina	
hares palays*	hare's palace	De palacio leporis	
heferne*	he-fern	De filice dicto os munda	**Dryopteris filix-mas**
hemlocke (OE)*	hemlock	De cicuta	Conium maculatum
hempe (OE)	hemp	De Canapis	Cannabis sativa
henbane (ME)*	henbane	De Iusquiamo	Hyoscyamus niger
herbe John (ME)	herb John	De Iperyco	Hypericum perforatum
/herbe/ paralysy		De herba paralisi / De primula veris	Primula veris / P. vulgaris
herbe phylyp*	herb Philip	De Saponarya	Saponaria officinalis
herbe squyn- antyke*		De herba squinancia	Asperula cynanchia
herbe of muske*	herb of musk	De Ima muscata	
herbe (grasse) of vine*	herb (grass) of vine	De herba squinancia	Asperula cynanchia
herbe of ynde*	herb of India	De Tetrahit	
Hercules grasse*	Hercules' grass	De stycados cytryne	Helychrysum stoechas
hertes tongue (ME)	hart's tongue	De stolopendria	Phyllitis (Asplenium) scolopendrium
hocke (OE)	hock	De malua ortulana	Althæa rosea
hogges meate*	hog's meat	De Uulfago	Cyclamen europæum
holyhocke (OE)	hollyhock	De malua ortulana	Althæa rosea
holy thystle	holy thistle	De Cardone benedicto	Cnicus benedictus
hondes tonge (OE)	hound's tongue	De cynoglossa	Cynoglossum officinale
honysocle (ME)*	honeysuckle	Coronaria	Melilotus
hoppes (ME)*	hop(s)	De lupulo	Humulus lupulus
horehounde (OE)	horehound	De Marrubium	Marrubium vulgare
horshele (OE)*	horseheal	De Enula campana	Inula helenium
horsmynte (ME)*	horsemint	De mentastro	Mentha longifolia
howsleke (ME)	house leek	De semper viua	Sempervivum tectorum
hye malowe*	high mallow	De Altea	Althæa officinalis
hygtaper*	hig(h) taper	De tapso barbato	Verbascum thapsus
hyndhele (OE)*	hindheal	De Ambrosiana	Teucrium scorodonia
jenepre (ME)*	juniper	De Iunipero	Juniperus communis
jōbarde (ME)*		De semper viua	Sempervivum tectorum
keyes	key	De lingua auis	Fraxinus excelsior (the seed)
kneholme (OE/ME)*	knee holm(e)	De genestula	Ruscus aculeatus

Name in GH	Modern form	Latin heading in GH	Modern scientific name
knotgrasse	knotgrass	De lingua passerina	Polygonum aviculare
knotwort (ME)	knotwort	[Coronaria]	Polygonum aviculare
kynges crowne	king's crown	De Melliloto	Melilotus
langdebefe (ME)*		De buglossa	Anchusa
lauendre (OE)	lavender	Spacea	Lavandula angustifolia
laurel (ME)	laurel	De lauro	Laurus nobilis
laureole (ME)		De laureola	Daphne laureola
leke (OE)	leek	De Porro	Allium porrum
lentyle (ME)	lentils	De lentico / De lentibus	Lens culinaris
lentylles of the water	lentils of the water	De lentycula aque	Lemna minor
lesse burre	lesser bur(r)	De lapaceola	Arctium
lesse consoulde (ME)	lesser consoud	De consolida minori	Bellis perennis
lesse morell (ME)	lesser morel	De solatro	Solanum nigrum
lesse mugwort	lesser mugwort	De Arthemesia minima	Artemisia
lesse saxifrage*	lesser saxifrage	De Saxifraga minori	Poterium sanguisorba / Sanguisorba minor
lesse skyrwit (see skyrwit)			
letuse (ME)	lettuce	De Lactuca	Lactuca sativa
longe plantayn(e)	long plantain	Lanceolata / De lanceolata	Plantago lanceolata
louage (ME)*	lovage	De leuistico	Levisticum officinale
lungwort (OE)*	lungwort	De pulmonaria	[a lichen]
lychwale (ME)	lichwale	De Milio solis	Lithospermum officinale
lychworte (ME)*	lichwort		Lithospermum officinale
lycoryce tre (ME)	liquorice	[De paracella]	Glycyrrhiza glabra
lylly (OE)	lily	De lilio	Lilium candidum
lymons (ME)	lemon	[De Balsamo]	Citrus limon
lyne (OE)	line	De femine lini	Linum usitatissimum
lyngwort*	lingwort	De eleboro albo	Veratrum album
lyons fote (OE)	lion's foot	De elleboro nigro / De pede leonis	Helleborus niger / Alchemilla
lyse grasse*	licegrass	De staphisagria	Delphinium staphisagria
lytell burre	little bur(r)	De lapaceola	Arctium
lytell clote*	little clot(e)	De ungula caballyna	Tussilago farfara
lytell plantayn*	little plantain	Lanceolata	Plantago lanceolata
lyuerwort (ME)*	liverwort	De epatica	Marchantia polymorpha
madder (OE)	madder	De Rubea	Rubia tinctorum
malowe(s) (OE)*	mallow	De Altea / De malua	Althæa / Malva
mandragora / mandrake (ME)	mandrake	De Mandragora	Mandragora officinarum
mares tayle (ME)*	mare's tail	De Iparis vel cauda equina	Equisetum (?)
Margetym gentyll*		De maiorana	Majorana hortensis (Origanum majorana)
mariorayne (ME)	majoram	[De Fistularia]	Majorana hortensis (Origanum majorana)
mary gowles (ME)	gules / marigold	Calendula	Calendula officinalis

Name in GH	Modern form	Latin heading in GH	Modern scientific name
maydenwede (ME)*	[mayweed]	Adianthos	Anthemis cotula
maydin here (ME)*	maidenhair	Capilli veneris	Adiantum capillus-veneris
maythen (OE)*	maithen (mathes / maithes)	Consolida media	Anthemis cotula
mederacle*	[mead rattle]	De camephiteos	Rhinanthus
medlers (ME)*	medlar	De Nespilis	Mespilus germanica
meke galyngale*		De Aristologia rotunda	Aristolochia rotunda
melisse (ME)		De melissa	Melissa officinalis
mellilot (ME)	melilot	De Melliloto	Melilotus
melons (ME)	melon	De Melonibus	Cucumis melo
mercury (ME)*	mercury	De mercuriali	Chenopodium bonus-henricus
mespile (ME)		De Nespilis	Mespilus germanica
meu	meu	De Meu	Meum athamanticum
moderwort (OE) / the moder of herbes (ME)	motherwort / mother of herbs	De Arthemesia	Artemisia vulgaris
molberyes (ME)	mulberry	De mora celsi	Morus nigra
moleyne (ME)	mullein	De tapso barbato	Verbascum thapsus
morell (ME)*	morel	[De lupulo]	Solanum nigrum (?)
more consoulde (ME)		De consolida maiori	Symphytum officinale
more morell (ME)		De solatro rustice	Atropa belladonna
mosse (OE)*	moss	Vsnea. vel muscus arborum	
mows eare (ME)	mouse-ear	De pilocella	Hieracium pilosella
mugwort (OE)	mugwort	De Arthemesia	Artemisia vulgaris
mussherons (ME)*	mushroom	De fungis	[fungi, edible]
mustarde (ME)*	mustard	De Eruca / De semine napei	Sinapis
myddle consoulde (ME)*	middle consoud	Consolida media	Anthemis cotula
myddle mugwort*	middle mugwort	De Arthemesia minor	
mylle (OE)	mile	De Milio	Panicum mileaceum
myllet (ME)	millet	[De Sizania]	Panicum mileaceum
myllefoyle (ME)*	milfoil	De millefolio	Achillea millefolium
mynte (OE)*	mint	De menta	Mentha spicata
myrte (ME)*	myrtle	De Mirto	Myrtus communis
nenufar (ME)*	nenuphar	De nenufare	Nuphar / Nymphæa
nespyte*		De calamento	Calamintha
nettle (OE)	nettle	De vrtica	Urtica
nutmygge (ME)	nutmeg	De nuce muscate	Myristica fragrans
nuttes of Inde (ME)*	nut of India (Indian nut)	De nuce Indica	Cocos nucifera
nyght shade (OE)	nightshade	De solatro	Solanum nigrum
oke(n)tre (OE)*	oak	Arbor glandis latine	Quercus robur
oke ferne (ME)*	oak fern	De polipodio	Polypodium vulgare
oleandre*	oleandre	De oleandro	Nerium oleander
olyues (ME)	olive	De Oliua	Olea europæa
onyon (ME)	onion	De Cepe	Allium cepa
open arses (OE)	open arse	De Nespilis	Mespilus germanica
orenges (ME)	orange	[De alcamia]	Citrus sinensis

Name in GH	Modern form	Latin heading in GH	Modern scientific name
orient saffron (saffron of orient)*	orient saffron	De croco	Crocus sativus
orygan (ME)*	organ	[De gelasia, De fungis, De faseolis]	Origanum vulgare (?)
otes (OE)*	oats	De Auena	Avena sativa
o[u]r ladyes saele [for seale] (ME)	lady's seal	De Sigillo sancte marie	Polygonatum multiflorum
oxtongue (OE)*	oxtongue	De buglossa	Anchusa
pagle (ME)*	paigle	De herba paralisi	Primula veris
panyke (ME)*	panic	De panico	Setaria italica
peches (ME)	peach	De persicis	Prunus persica
pedelyon (ME)		De elleboro nigro / De pede leonis	Heleborus niger / Alchemilla
pellyter (ME)*	pellitory	De serpyllo	Thymus serpyllum (?)
peleter of Spayne (ME)*	pellitory of Spain	De eleboro albo	Veratrum album
peny wort (ME)*	pennywort	Ameos	
peper (OE)*	pepper	De pipere	Piper nigrum
percely (ME)*	parsley	De petrocilio	Petroselinum crispum
percely of Macedony	parsley of Macedonia	De petrocilio macedonico	Smyrnium olusatrum
peres (OE)	pear	Pira	Pyrus communis
perwynke (OE)*	periwinkle	De prouinca	Vinca minor
petymorell (ME)		De solatro	Solanum nigrum
pimpinell / pympernell (ME)*	pimpinell / pimpernell	De pimpinella	Pimpinella saxifraga
pireter (ME)*		De piretro	Parietaria diffusa
plantayne (ME)	plantain	De plantagine	Plantago major
plommes (OE)	plum	De prunis	Prunus domestica
policary (ME)*		De policaria	Pulicaria
politryke	polytrich (OED)	De polytryco	Asplenium
polipodi (ME)	polypody	De polipodio	Polypodium vulgare
pomgarnades (ME)*	pomegranate	De balaustia / De granatis	Punica granatum
pompon*	pompion	De Cucumero	Cucumis melo
poppy (OE)	poppy	De Papauere	Papaver
porcelayne (ME)*	purslane	Dc portulace	Portulaca oleracea
prestes hode*	priest's hood	De Iaro	Arum maculatum
prymerolles (ME)	primerole	De primula veris	Primula vulgaris
pyllulary*		De staphisagria	Delphinium stafisagria
pyne apples (ME)*	pine apple	De pineis	Pinus (the fruit)
pyne tree (OE)	pine	De pineis	Pinus
pyony (OE)*	peony	De peonia	Pæonia
quekes (OE)	quick(s)	De gramine	Agropyron repens
quynces (ME)	quince	De citoniis	Cydonia oblonga
quynce apples	quince apple	De citoniis	Cydonia oblonga (the fruit)
radysshe (OE)	radish	De radice	Raphanus sativus
rampsons (OE)	ramsons	Allium latine	Allium ursinum
rapes (ME)	rape	De rapa	Brassica rapa

80

Name in GH	Modern form	Latin heading in GH	Modern scientific name
rasyns of carans (ME)*		Uua passe	
ray (ME)	ray	De Sizania / Zizania	Agrostemma / Lolium temulentum
rede (ME)*	reed	De canna	Phragmites communis
reed brere (ME)*	red briar	De tribulo marino	Eryngium maritimum
reed docke (ME)	red dock	De Lapacio	Rumex
reed mader (ME)*	red madder	Aristologia longa latine	Aristolochia longa
reed poppy (ME)	red poppy	De Papauere	Papaver rhoeas
remcope*		Morsus diaboli	Succisa pratensis
rewbarbe (ME)*	rhubarb	De Reubarbaro	Rheum
rose (OE)	rose	De Rosa	Rosa
rosmary (ME)	rosemary	De rose marino	Rosmarinus officinalis
ruddes (ME)*	ruds	Calendula	Calendula officinalis
rue (ME)*	rue	De ruta	Ruta graveolens
rue of the felde (OE)*	field rue	De pigano	Ruta montana
ryb(b)wort (ME)*	ribwort	De laureola	Plantago lanceolata
rye (OE)	rye	De Siligo	Secale cereale
rys (ME)	rice	De Riso	Oryza sativa
rysshe (OE)	rush	[De calamo aromatico]	Cyperus/Juncus
saffron (ME)*	saffron	De Croco	Crocus sativus
Salamons seale	Solomon's seal	De Sigillo sancte marie	Polygonatum multiflorum
sanguinary (ME)	sanguinary	De Persicaria / De sanguina[r]ia / De bursa pastoris	Polygonum hydropiper / Digitaria sanguinalis / Achillea millefolium / Capsella bursa-pastoris
saponary (ME)	saponary	De Saponarya	Saponaria officinalis
sapyn (tre) (ME)*	sapin	De terbentine	
sarazyns mynt*	Saracen's mint	De menta	Mentha gentilis (?)
satirion (ME)	satyrion	De palma cristi / De satirione	[some orchid]
sauerey (ME)	savory	De satureia	Satureja hortensis
sauyn (OE)*	savin(e)	De Sauina	Juniperus sabina
sawge (ME)	sage	De Saluia	Salvia officinalis
saxifrage (ME)*	saxifrage	De Saxifraga	Ceterach officinarum (?)
saynt Johannis (Johns) wort (ME)	St. John's wort	De Iperyco	Hypericum perforatum
saynt peterworte*	St. Peter's wort	De primula veris	Primula vulgaris
scabwort (ME)	scabwort	De Enula campana	Inula helenium
scabyous (ME)*	scabious	De scabiosa	Centaurea scabiosa (?)
scaryole (ME)*		[De endiuia / De Boragine]	Cichorium endivia (?)/ Lactuca serriola
see onyon*	sea onion	De squilla	Urginea maritima
selfegrene (OE)*	selgreen / sengreen	De semper viua	Sempervivum tectorum
selfe heale (ME)	selfheal	De pimpinella	Pimpinella saxifraga
senacion*		De narsturcio	
sene (ME)	senna	De Sene	Cassia acutifolia
senechon (ME)*		De senacionibus	Senecio vulgaris
seneuy (ME)*	senvy	De semine napei	Sinapis

Name in GH	Modern form	Latin heading in GH	Modern scientific name
sentynode (ME)	centinody / centinode	De lingua passerina	Polygonum aviculare
s[p]ereworde (ME)*	spearwort	De flamula	Ranunculus flammula
serpentyne / of dragons / (ME)	serpentine	De serpentina	Dracunculus vulgaris
setwall (ME)*	setwall	De Fenegreco / De Zedoare	Trigonella foenum-græcum / Curcuma zedoaria
shepeherds purs (ME)	shepherd's purse	De bursa pastoris	Capsella bursa-pastoris
skyrwyt (ME)*	skirret	De baucia / De Eruca	Pastinaca sativa / Eruca sativa
sloe (OE)*	sloe	De Achasio	Prunus spinosa
smalache (ME)	smallage	De Apio	Apium graveolens
small mugwort	small mugwort	De Arthemesia	Artemisia
smerewort (OE)*	smearwort	De Aristologia rotunda	Aristolochia rotunda
smoke of the erthe	smoke of the earth (earth smoke)	De fumo terre	Fumaria officinalis
snakegrass*	snakegrass	De serpentina	Dracunculus vulgaris
sorbes (ME)*	sorb	De Sorbis	Sorbus domestica
sorell (ME)	sorrel	Acetosum latine	Rumex acetosa
sorell de boys (ME)*		[De Agarico]	Oxalis acetosella
sowthistle (ME)*	sow-thistle	De Cardone benedicto	Cnicus benedictus
sparow tongue (ME)	sparrow tongue	De lingua passerina	Polygonum aviculare
sperage (ME)*	sperage	De sperago	Asparagus officinalis
spewynge nuttes	spewing nut	De nuce vomyca	Strychnos nux vomica
spourge (ME)*	spurge	De catapucia	Euphorbia
spyke / spyknarde (ME)*	spike / spikenard	De spicnardo	**Nardostachys jatamansi**
spyke celtyk		De spicnardo	Valeriana celtica
spynache (ME)	spinach	De spinachia	Spinacia oleracea
squinant (ME)		Squinanto	Cymbopogon schoenanthus
squyll (ME)	squill	De squilla	Urginea maritima
stammarche (OE)*	stanmarch	De Apio / De Petrocilio macedonico	Apium graveolens / Smyrnium olusatrum
s[t]rawberye(s) (OE)*	strawberry	De Fragaria	Fragaria vesca
strofulary*		De strofularia	
styche wort (OE)*	sti(t)chwort	De lingua anseris	Stellaria holostea (?)
sugre rede	sugar reed (cane)	De canna mellis	Saccharum officinarum
swete almondes (ME)	sweet almond	Amigdala latine	Prunus Amygdalus (v. dulcis)
swete commyn (ME)*	sweet cumin	Anisum latine & grece	Pimpinella anisum
swynefenell (ME)*	swine fennel	De peucedano	Peucedanum officinale
swynes brede*	swine bread	[De Apio]	Cyclamen europæum
swynes grasse (OE)*	swine's grass	De lingua passerina	Polygonum aviculare
synkefoyle (ME)*	cinquefoil	**De Penthafilone**	Potentilla reptans
tamaryte*	tamarisk	De tamaristo	**Tamarix gallica**
tame cresses*		**De Narsturcio**	Lepidium sativum
tame garlike (ME)		Allium latine	Allium sativum
tame malowe (ME)		De Altea / De Malua	Althæa rosea (?)
tame pere (ME)		Pira	Pyrus communis
tansey (ME)	tansy	[De Achasio]	Tanacetum vulgare

Name in GH	Modern form	Latin heading in GH	Modern scientific name
tapsebarbe*		De tapso barbato	Verbascum thapsus
tasyll (OE)*	teasel	De tribulo marino	Eryngium maritimum
thre cornerde rysshe	three-cornered rush	De Cypero	Cyperus longus
thre (.iii.) leued grasse (ME)	three-leaved grass	De trifolio	Trifolium
thystle (OE)	thistle	[De radice yringorum / De tribulo marino]	Carduus, Cirsium, etc.
thystle of the see (ME)	sea-thistle	De tribulu marino	Eryngium maritimum
tintymall*	tithymall	De Tyntymallo	Euphorbia
tyntymall of babylon		De Tyntymallo	Euphorbia
tode stoles (ME)	toadstool	De Ipoquistidis	[fungi, non-edible]
tormentyll (ME)*	tormentil	De tormentilla / De Albatra	Potentilla erecta
trefle (ME)	trefoil	De trifolio	Trifolium
tutson (ME)*	tutsan	De Agno casto	Vitex agnus-castus
valerian (OE)*	valerian	De valeriana / De silfu	Valeriana / Polemonium cæruleum
veruayne (ME)	vervain	De Berbena	Verbena officinalis
vyolettes (ME)	violet	De violis	Viola odorata
walfarne*	wallfern	De polytryco	Asplenium
wall nuttes (OE)	wallnut	De nuce communi	Juglans regia (the fruit)
walworde (OE)*	wallwort	De Ebulo / De piretro	Sambucus ebulus / Parietaria diffusa
warence (ME)		De rubea / De spergula	Rubia tinctorum
wartwort (ME)*	wartwort	De herucaria	
water cress (ME)	water cress	De Narsturcio	Rorippa
water flagge*	water flag	Acorus	Iris pseudacorus
water onion	water onion	De Squilla	Urginea maritima
weybrede (OE)	waybre(a)d	De plantagine	Plantago major
wheate (OE)	wheat	De frumento	Triticum æstivum
white poppy (OE)	white poppy	De Papauere	Papaver somniferum
whyte bryony	white bryony	De brionia	Bryonia alba
whyte elebore (ME)*	white hellebore	De eleboro albo	Veratrum album
whyte vyne*	white vine	[De lupulo]	Bryonia alba (?)
wolfe thystle (ME)*	wolf thistle	De cameleonta	Carlina
woodbynde (OE)	woodbine	De caprifolio / De matri-silua / De Uolubilis	Convolvulus / Lonicera
wood crabbes (ME)	wood crab	De macianis pomis	Malus sylvestris
woodnep*		Ameos	
woodroue (OE)*	woodruff	De astula regia	Asphodelus / Galium odoratum
woodsorell*	woodsorrel	De Alleluya	Oxalis acetosella
woodyp*		De genestula	
wormwood (ME)	wormwood	Absinthium latine	Artemisia absinthium
wylde aloes*	wild aloe	De ligno Aloes	
wylde apples	wild apple	De macianis pomis	Malus sylvestris
wylde blacke beryes	wild blackberry	De mora celsi	Rubus fruticosus

Name in GH	Modern form	Latin heading in GH	Modern scientific name
wylde bourache (ME)*	wild borage	De buglossa	Anchusa
wylde cawles (ME)*	wild cole	De Eruca	Brassica/Sinapis
wylde cowcomers (ME)*	wild cucumber	De Electerio	Ecballium elaterium
wylde cresses (ME)*	wild cress	De Narsturcio agresti	
wylde galyngale*	wild galingale		Cyperus longus
wylde garlyke (ME)*	wild garlick	Allium latine / Scordeon	Allium ursinum / A. vineale
wylde gourde(s) (ME)	wild gourd	De brionia / Colloquintida	Bryonia dioica / Citrullus colocynthis
wylde hempe (ME)*	wild hemp	De Canapis	Eupatorium cannabinum
wylde leke	wild leek	De Porro	Allium
wylde letuse (ME)*	wild lettuce	De lactuca siluestri / De lactuca agresti	Lactuca serriola (?)
wylde malowes (ME)*	wild mallow	De maluauisco	Althæa officinalis (?)
wylde molberyes	wild mulberry	De mora celsi	Rubus fruticosus
wylde mynte (ME)*	wild mint	De mentastro	Mentha longifolia
wylde neppe (ME)*	wild nep	De brionia	Bryonia dioica
wylde percely (ME)*	wild parsley	De Sinomo	Sison amomum
wylde peres (ME)	wild pear	Pira	Pyrus
wylde poppy (ME)	wild poppy	De Papauere	Papaver rhoeas
wylde rapes*	wild rape	De Rapiastro	Brassica / Sinapis
wylde rue (ME)*	wild rue	De piganio	Ruta montana
wylde sawge (ME)*	wild sage	De eupatorio	Teucrium scorodonia
wylde smalache*	wild smallage	De Apio ramio	
wylde tasyll (ME)*	wild teasel	De virga pastoris	Dipsacus fullonum ssp. fullonum
wylde tyme (ME)*	wild thyme	De polio montano	Thymus serpyllum (?)
wylde valeryan*	wild valerian	De silfu	
wylde wyne (ME)*	wild vine	De Lambrusca	Vitis labrusca
wyldynges	wilding	De macianis pomis	Malus sylvestris
wyloue (tree) (OE)	willow	De salicibus	Salix
wytmynt (ME)*	white mint	De menta romana	Mentha gentilis (?)
yarowe (OE)	yarrow	De mille folio / De sanguina[r]ia	Achillea millefolium
yelowe flagge*	yellow flag	De spatula fetida	Iris pseudacorus
yringe (ME)	[eryngo]	De radice yringorum	Eryngium maritimum
ysope (OE)	hyssop	De ysopo	Hyssopus officinalis
yuy (OE)	ivy	De edera magna	Hedera helix

aaron	For identification of the plant as *Arum italicum* in *Circa instans*, see Camus 1886, p. 76.
ache	"Some call it botracium ... other herba scelerata." As *Ranunculus sceleratus* first recorded in GH (also in Turner 1538). As *Apium* ME.
affodylly	*Affodill(e)*, which is also in GH, is ME. In Turner 1538 "Affadyll et Daffadilly". The picture in GH (as in *Le grant herbier*) represents an iris, but the text describes a species of *Asphodelus*, possibly *A. lutea* (cf. Wessén 1924, p. 64).
agryotes	Cf. *damacenes*.
alcamet	For identification, see Camus 1886, p. 32, Wölfel 1939, p. 125 (note 159) and MED.
alleluya	For identification of the plant as *Oxalis corniculata* in *Circa instans* and *Le grant herbier*, see Camus 1886, p. 32, and Fischer 1929, p. 277.
aloe(n)	Cf. the text: "Aloe is made of the iuce of an herbe named Aloen ..." and the "Registre": "Aloe / a iuce so named". "Aloes a wood so named".
alysamder	For GH forms, see pp. 42 f. Now invariably in the plural.
anys	Variant GH spellings: *annys, an(n)es*.
artetyke	Lat. *herba arthritica*. The plant was used against the gout. Cf. Nordhagen 1954, p. 42.
baldymony	The plant the translator had in mind was probably *Gentianella amarella* (as with *felwort*; see below). In Banckes's herbal "Felworte or Baldomoyne". Cf. *gencyan*.
barley	For GH forms, see p. 42.
bearefote	Occasionally *beres fote*. Probably, like *beares bough / twygge*, for *Acanthus mollis*. Cf. Brodin 1950, p. 255, Camus 1886, pp. 30 and 43, and Fischer 1929, p. 271. In Turner 1538 for *Acanthus mollis* and *Helleborus niger*; in Turner 1548 only for *Helleborus*.
berberies	"Berberyes ben fruytes so named."
bethony	Also spelt *bethoni* and *betony* in GH. Cf. p. 34, for the name as recorded in OE.
blacke bryony	For identification, see OED and Britten & Holland, p. 69.
blacke yuy	Cf. Britten & Holland, p. 276.
blewe flourdelyce	"Iris hath a blewyishe reed floure". Also in VBD.
blodworte	Variant GH spelling: *blodeworte*. In ME only for *Capsella bursa-pastoris*.
bonewort	Probably for *Bellis perennis* (as in Turner 1538). See Brodin 1940, p. 255, and MED bon-wort. Cf. Bierbaumer 1979, pp. 16 f. Also as *benworte* in GH (sub "De Aloe"); cf. Britten & Holland, pp. 25 and 39.
borage	For GH forms, see p. 43.
brere	Sometimes paralleled with *thorn:* "Bedegart is a thorne or brere". Cf. Fries 1904, p.44.

brotherworte	Probably for *Origanum vulgare* (cf. MED brother-wort (c)).
bryght	Translation of Fr. *esclere*, i.e. *esclaire*, referring to *Chelidonium majus* or *Ranunculus ficaria* ("petite esclaire"). See Gerth van Wijk 1911, pp. 292 and 1124. Cf. also Marzell 3 (1977), 1252. Identified by Britten & Holland (p. 65) as *Ranunculus ficaria*, for which *bright-eye* has been recorded in dialects (Britten & Holland, p. 515, and Wright 1 (1898), p. 402).
buckesshorne	In the 1539 and 1561 eds. *buckeshorne*. Possibly *Coronopus squamatus*. Cf. MED and Brodin 1950, pp. 256 and 277. In Banckes's herbal "Lingua hyrcina" is "Buckeshorne or Swynykarce" (i.e. swine cress). Cf. also Franckenius 1659: "Coronopus ... Hiortehorn" (i.e. buck's horn).
buglosse	By the old herbalists the name *bugloss* was given to several plants with rough tongue-like leaves, especially to species of *Anchusa* and *Echium* (cf. Marzell 2 (1972), 183). For GH names denoting *Anchusa* or similar plants with rough leaves, see p. 50. *Buglosse* is also mentioned sub "De eufragia". Cf. Fischer 1929 (p. 269), who gives *bugula* as an old name for *Euphrasia*.
burit	Cf. Marzell 4 (1979), 105.
b[l]ynde nettell	*Bynde nettell* in all eds.
byrten tre	See p. 71. In the text *byrthen tre*.
calamynt	Variant GH forms: *calament, calomynt*. There is also a GH *calamynt of the mountayne* (*Calamintha officinalis*?) as a reflex of the Latin and French texts (see Camus 1886, p. 46). Cf. Marzell 4 (1979), 105.
caltrappe	For early forms, see MED calketrappe. Now usually in the plural. In ME applied to several plants (see MED). In 16th cent. herbals usually for *Centaurea calcitrapa*.
calues fote	Britten & Holland (p. 82) also assigns (erroneously) this name to *Allium vineale* in GH.
camelles strawe	For identification, se Camus 1886, p. 122. Cf. Gerard 1597 (p. 40), for "Scænanthus": "in English Camels haie, and Squinant". In Franckenius 1659 *Kameelhöö* 'camel's hay' for the same plant (cf. Marzell 1 (1943), 266). In Lyte 1578 for *Juncus effusus* (see OED and Britten & Holland, p. 83).
canne	Bot. ref. here obscure.
capparis	Variant GH form: *cappres*. In the 1539 and 1561 eds. *capars* and *capris*, respectively.
caule wortes	Variant GH spellings: *c(o)ole(s) wortes*. "There ben wynter caules / and somer caules".
celendyne	Cf. *bryght* and Britten & Holland, p. 95.
ceterach	Cf. Turner 1568: "I haue harde no English name of this Herbe / but it maye well be called in English Ceterache / or Miltwaste / or Finger ferne: because it is no longer then a mannes finger: or Scaleferne / because it is all full of scales on the innersyde".

cheruell	See pp. 39 and 43.
churles tryacle	"Yf it be eaten it putteth venym out of the body / and therfore it is called churles tryacle".
chyches	In GH 1561 *fyches.*
chycory	Variant GH form: *cicore* (sub "De lactuca leporina").
clarey	See p. 33. Cf. Marzell 4 (1979), 54.
clote	"A clote that bereth burres". For *bur* as plant name in GH, see *grete burre* and *lytell burre.*
clyuers(s)	OE *clife* for *Arctium lappa* and *Galium aparine* (Bierbaumer 1979, p. 52). In ME only attested for *Galium* (MED). In GH the pl. form only for *Galium.*
cokyll	For GH forms, see p. 42.
colloquintide	Alternates in GH with *colloquintida*. Also in *apple of colloquintide.*
comfrey	Variant GH form: *confrey* (cf. p. 43).
common + noun	See list at the end of the Overall list.
comyn	Variant GH form: *commyne.*
consoulde	In the Middle Ages this name was assigned to several plants on account of their consolidating (healing) virtues, in particular to *Symphytum officinale* (see Britten & Holland, p. 116). Cf. also Morton 1981, p. 113.
/less/ coronary	Identity obscure (cf. Camus 1886, p. 58).
cowslyp	In the "Registre" misspelt as *crowslyp*. VBD: *cowslop.*
crayfery	See p. 41.
cress	The *cresses* of the heading refers to various kinds of "cresses". As simplex (sg.) usually for *Lepidium sativum* in GH.
croyt marine	See p. 40.
cukowes meate	In the "Registre" *cocowes meate.*
cynamone	For GH forms, see p. 42 (also *synamon/synamum*).
cytrons	ME *citrine* (MED). For a possible ME record, see Wilson 1979 (p. 506).
damacenes	"Those [cheryes] with that bytternesse ben called damacenes and the other agryotes." Usually in English synonymous with *damson*. The early eds. of *Le grant herbier* (incl. the *Arbolayre*) have *amarenes*, whereas the 16th cent. eds. of *Le grant herbier* seen by me have *damacenes* (like GH).
darnell	The name—now usually applied to *Lolium temulentum*—was given by the early herbalists to several plants (see e.g. Prior 1863, pp. 63 f.). The picture in GH shows *Agrostemma* (no picture in *Le grant herbier*). In the text it is said that "Zizania is an euyll wede y^t groweth in the wheate / and corrupteth whan the weder is drye". In VBD for *Agrostemma.*
date of Ynde	Cf. OED tamarind.
dodyr	The reference in GH is usually to *Cuscuta epilinum* (e.g.: "Cuscuta ... is an herbe y^t wyndeth about flax or lyne growynge"). Cf. VBD ("Dodyr whiche groweth in the flaxce") and Wessén 1924, p. 68.

dogfenell	For identification, see Camus 1886, p. 104, and Wölfel 1939, p. 139 (note 653). Britten & Holland (p. 155) suggests *Peucedanum palustre*. In ME usually for *Anthemis cotula* (see MED and Brodin 1950, pp. 252 and 262). Cf. *mydde consolde (Anthemis cotula)* given as GH equivalent of *dogfenell* sub "De peucedano".
dogges rose	For *Rosa canina* first recorded in Gerard 1597.
doues fote	By Camus 1886 (p. 102) identified as *Geranium columbinum*. In *Agnus Castus* (c. 1440) and in Turner 1548 for *G. molle* (see Brodin 1950, p. 285, and Turner 1965, p. 241, respectively).
drawke	Cf. sub *darnel* ("Zizania"). Cf. Britten & Holland (p. 159) where identified as *Agrostemma*: "*Lychnis Githago* is called *Drawk* in the Grete Herball". For the formal variation of the word, see Britten & Holland, *loc. cit.*
ducke meate	In the "Registre" *duckes meate*.
dytany	Variant GH forms: *dyptan, dyptany* (cf. p. 43). VBD: *dyptan, dytteyn*. Probably identical with Turner's *righte dittany* (see Turner 1548 and 1965). Cf. also Frisk 1949, p. 229.
egrymony	Variant GH form: *agrymony*.
elebore	GH *elebore* (without modification) usually refers to *Veratrum album*, occasionally to *Helleborus niger*, as sub "De Antimonio": "The powdre of elebore that is pedellion".
erththought	See p. 38.
eufrace	Variant GH form: *eufragye* (see p. 43). In Banckes's herbal *eufrasy*. In VBD *eufrasie* and *iyen comfort* (< German *Augentrost*). In GH 1561 *eufrace* and *eye brighte*.
eupatory	Cf. for identification: "the wylde [sawge] yᵗ is called eupatory" (De eupatorio) and "Eupatorium is an herbe otherwyse called Saluia agrestis" (ibid.).
felwort	Cf. *baldymony*.
fenegreke	Variant GH forms: *fenugrec, fenygreke*.
flag	For identification, cf. Turner 1538 (1965). Cf. VBD: "Gladiolus flag" (*Iris pseudacorus*).
fumyterry	Variant GH forms: *fume terri, fumoterre, fumyterre*. In Banckes's herbal *fumytory*.
fyrre	Used as equivalent of *sapyn* (which see).
fyue leued /grasse/	*Fīflēafe* is OE, *fiveleaved grass* ME (Brodin 1950, p. 267). In the "Registre", only the composite with *grass* (as in GH 1561).
galles nuttes	"the fruite of okes". Also in GH as *galle nuttes* and *nutte of gall / nutgalle*.
gangelon	See p. 41.
gardyn gynger	Cf. *dytanty*.
gardyn malowes	Although in GH given as a synonym of *holyhocke* ("De malua ortulana. Holyhocke."), the text rather indicates *Lavatera arborea*: "It is a grete malowe in maner of a tre with grete leues". Cf. the current name *tree mallow*. For identification, see Camus 1886, p. 88.

gardyn mynte	First attested in Banckes's herbal (1525), where paralleled with *howse mynte*.
gardyn saffron	Cf. sub *saffron*.
garlyke	Misspelt as *gorlyke* in the heading.
gencyan	Probably *Gentiana lutea* is intended (as in Turner 1548). Cf. OED and Camus 1886, p. 71.
germaundre	In the "Registre" *garmaundre*.
goosbyll	The plant implied in the *Lingua anseris* or *bec doye* of *Le grant herbier* is presumably *Potentilla anserina* (Camus 1886, p. 85), as hinted at in the text: "the leues ben lyke the leues of ferne". The translator may have mixed up *Lingua anseris* with *Lingua avis* (i.e. stichwort). Cf. Marzell 4 (1979), 491, and Mowat 1887, p. 103. In GH *goosbyll* and *styche wort* are given as equivalents. Cf. Britten & Holland, p. 213.
goosfote	Not in this sense in OED. Also called *sanguinary* in GH. For identification, see Camus 1886, p. 113, Fischer 1929, p. 267 and Marzell 3 (1977), 533. In Turner 1548, as in the current nomenclature, for *Chenopodium*.
grenes	ME for 'green plants'.
grete centory	Cf. Britten & Holland, p. 96.
grete plantayne	Also in VBD. ME *(the) more plantain.*
gromyll	Variant GH form: *gromell*.
grownswell	VBD: *groundswell*. For the etymology of the name, see Bierbaumer 1975–79 and Kärre 1924.
gylofre	Variant GH form: *geloffre*. For identification, cf. MED gilofre.
gynger	Variant GH form: *genger*.
hare ballockes	Also in Turner's herbal. ME *ballock wort* (MED). In Lyte (1578) *ballock grasse*.
hare berde	Also in VBD.
hare trefle	Cf. Camus 1886, p. 126.
hares letuse	Cf. Camus 1886, p. 80, and Britten & Holland (p. 242), where the plant is identified as *Sonchus oleraceus* (as in OED). In the text the plant is described as having "leues lyke cicore" (i.e. chicory).
hares palays	VBD: *hare castell*. Identified by Camus 1886 (p. 100) and Fischer 1929 (p. 108) as *Asparagus tenuifolius*. The plant is said to be "leued lyke fenell . . . and it bereth no floure/but a reed bery". By OED and Britten & Holland taken to be a GH equivalent of *hares letuse*. Cf. Fischer 1929 (*loc. cit.*): "Die Südländer verstehen darunter [i.e. "Palatium leporis"] *Asparagus tenuifolius*, die Nordländer eine grossblättrige Komposite wie Crepis, Sonchus".
heferne	"Filex masculus is heferne". Cf. the modern name *male fern.*
hemlocke	Variant GH form: *hemloke*.
henbane	For GH forms, see p. 43. VBD: *henquale* (not in OED).
herbe phylyp	Lat. *herba Philippi*. Cf. Marzell 4 (1979), 105.
herbe squynantyke	See Britten & Holland, p. 259.

herbe of muske	Identity obscure.
herbe (grasse) of vine	For identification, see Camus 1886, p. 75.
herbe of ynde	Identification uncertain. Cf. Camus 1886, p. 125, Fischer 1929, p. 21, and Wölfel 1939, p. 114.
Hercules grasse	For identification, see Camus 1886, p. 123, Fischer 1929, p. 108, and Marzell 2 (1972), 789.
hogges meate	For identification, see Fischer 1929, p. 266. Not in OED for *Cyclamen.* Cf. *swynes brede.*
honysocle	Discrepancy between GH text and English name in the heading. Cf. VBD *honysocle* ("Corona regia") and *kynges crowne* below.
hoppes	In the "Registre" *hoppe.* For the ME provenance of *hop(s),* see p. 34.
horshele	Cf. p. 33.
horsmynte	Equivalent of *wylde mynte* (which see). For identification, see Camus 1886, p. 91, and Fischer 1929, p. 272, and cf. Turner 1548 (1965).
hye malowe	"Altea is a hye Malowe." In the "Registre": "Altea / hye malowe". The plant is said to grow "in moyste places and feldes". Cf. *wylde malowe.* Cf. Franckenius 1659 and Wessén 1924, recording *hög kattost* 'high mallow' for *Althæa officinalis* and *Malva sylvestris* (Franckenius).
hygtaper	In the "Registre" *hyghtapper.* The name is first recorded in GH. VBD: *hygh taper/hie tapers.* Turner 1548 (first OED record): *higgis taper.* Lyte (1578) and Gerard (1597): *high-taper.* For the formal variation of the name, see OED *hagtaper, hightaper,* and Britten & Holland, pp. 238 and 260. Cf. also Prior 1863, pp. 112 f. The forms in *high,* which are considerably earlier than previously known (OED first record from 1605), may be original or folketymological adaptations of *hig* (of uncertain origin). The dried spikes of the plant were used for torches: "[it] bereth a lōge stalke whereof is made a maner of taper or lynke yf it be talowed".
hyndhele	For the etymology and bot. ref. of this name, see Nordhagen 1965, especially pp. 79 ff. Cf. also Bierbaumer 1979, p. 135, and Marzell 4 (1979), 675. The name is last recorded in Gerard 1597: "Hyndhæle is *Ambrosia*". Cf. Turner's (1548) *ambrose* for the same plant. GH equivalents: *eupatory, wylde sawge.*
jenepre	For GH form, cf. OED *juniper* and Britten & Holland, p. 278. VBD: *jeneper, genyper.*
jōbarde	Britten & Holland (p. 280) gives incorrectly *jo-barbe* as the GH form.
kneholme	See p. 33.
langdebefe	Cf. p. 34 and *buglosse.*
lesse saxifrage	In the text it is stated that "Sorba stella [i.e. sorbastrella] ... is lyke pympernell". Cf. Fischer 1929, pp. 280 and 283 and Marzell 4 (1979), 83. The name does not refer to *Saxifraga* (as suggested in OED *saxifrage*).

louage	For identification, see Britten & Holland, p. 315.
lungwort	The text describes a species of *Pulmonaria*, but the picture represents a lichen. Cf. Camus 1886, p. 108, Le Strange 1977, p. 212, Britten & Holland, p. 317 and Wessén 1924, p. 79.
lychworte	In GH only in the "Registre", where used as equivalent of *gromyll:* "Gromyll milium solis / lychworte". For forms and bot. ref. of the name in ME, see MED lith-wort.
lyngwort	Cf. Turner 1538 (*lyngwort* = "Elleborum Album"). See also sub *peleter of Spayne*.
lyse grasse	Cf. Britten & Holland sub *lousewort* and Brodin 1950, pp. 246 f.
lytell clote	Cf. the 1561 ed.: "De Ungula caballina. Litell clote. Tussilago, bulfote, horse hofe." *Lytell clote* ("vngula caballina") is also in VBD.
lytell plantayn	VBD: *small plantayne*. ME (*the*) *less plantain*.
lyuerwort	Variant GH form: *leuerwort*.
malowe(s)	The singular form appears with "De Altea".
mares tayle	Cf. Fischer 1929, p. 268, and Marzell 2 (1972), 234. Today *mare's tail* usually denotes *Hippuris*.
Margetym gentyll	For GH forms, see p. 40. In the "Registre" *gentyll margetyn*.
maydenwede	Not equivalent to *maidenhair* (*Adiantum*) as suggested in OED maidenweed. Cf. the text: "the herbe hyght [i.e. called] politricum is an other herbe / this herbe adianthos hath leues lyke to coryandre".
maydin here	In the "Registre" *maydē here*.
maythen	Cf. Britten & Holland, p. 327.
mederacle	See pp. 41 f. VBD: *medracle*.
medlers	In the "Registre" "mydlers or nefles" (cf. p. 51).
meke galyngale	First attested in Banckes's herbal (1525).
mercury	GH variant: *mercuryall* (Le grant herbier: *mercuriale*). For identification, cf. Fischer 1929, p. 264. Cf. also MED mercurie ("prob. chenopodium"). In Turner 1965 identified as *Mercurialis perennis* (cf., however, Britten & Holland, p. 332).
morell	Possibly here equivalent to *petymorel* (cf. Britten & Holland, p. 341, Banckes's herbal: "Morell or Nyght shadowe", and Turner 1538: "Morel, aut nyghtshad"). The reference may also be to *Atropa belladonna*, the "more morell" (see below).
mosse	"Vsnea is of dyuers maners". Cf. Fischer 1929, p. 287 ("Usnea barbata"), and Wessén 1924, p. 84.
mussherons	Cf. p. 23.
mustarde	In GH only in *mustarde* (*musterde*) *sede*. VBD: *mostarde herbe/sede*.
myddle consoulde	In GH used as equivalent of *maythen*. Cf. sub *dogfenell*. The *Consolida media* of *Circa instans* = *Symphytum* (Camus 1886, p. 56).

myddle mugwort	Identity obscure. Cf. Camus 1886, p. 38 ("Tanacetum vulgare").
myllefoyle	For GH forms, see p. 43.
mynte	The *myntes* of the heading refers to various kinds of "mints".
myrte	The variant *myrtylle* denotes the fruit: "Myrte is a lytell tre so called / ye which tre bereth a fruite that is named Myrtylles."
nenufar	"It is of two maners. One is whyte / and another yelowe." In Turner 1538 called *water rose.* Cf. Turner 1568: "Nymphea is named of the apothecaries nunefar [for nenufar]/ in Englishe water rose or water lili." Lyte 1578 and Gerard 1597 have "White water Lillie" / "Yellow water Lillie".
nespyte	Cf. Britten & Holland, p. 353.
nuttes of Inde	Cf. Camus 1886, p. 96, and Wölfel 1939, p. 138 (note 623).
oke ferne	See Britten & Holland, p. 180. In Turner 1548 sub "Dryopteris".
oke(n) tree	GH *oker* 'acorn' is not in OED ("The okers eaten stoppeth the longue and excessyfe flux in women"). Cf. Britten & Holland, pp. 9 and 506.
oleandre	In the "Registre" *oliandre.*
orient saffron	Cf. sub *saffron.*
orygan	For identification, cf. Turner 1548 (1965).
otes	In the "Registre" *ote.*
oxtongue	Cf. sub *buglosse.*
pagle	OED first record from Fitzherbert 1530 (*pagyll*). Turner 1548: *pagle.* Tusser 1573: *paggles* (OED). The modern form (*paigle*) first attested in Gerard 1597. MED records *plaggis* from a Stockholm MS of ca 1450: "Tak kousloppes, þat is plaggis [read: pagglis], and primerose-leues, …". On this MS, see Brodin 1950, pp. 31 and 56, and Nordhagen 1954, p. 62. On the derivation and semantics of *paigle,* see Nordhagen 1954, pp. 62 ff. (where associated with OE *paegel* 'wine vessel, gill' and Middle Dutch *pegel* 'gauge, scale, mark' as referring to the structure of the (usually heterostylous) flowers). Another possible explanation of *paigle* would be to connect it with *paggle* 'bulge' as referring to the inflated calyx (the motive would then be the same as that underlying *slip/slop* in *cowslip/cowslop* as suggested in Nordhagen 1954).
panyke	Cf. Turner 1568: "But it hathe no name in English yet, but it may be called panick after y^e Latin."
pellyter	"It grypeth and spredeth on the erthe." For GH forms, see p. 43. For identification, see Camus 1886, p. 117, Fischer 1929, p. 286, Marzell 4 (1979), 700, and OED pelleter 2.
peleter of Spayne	Cf. Banckes's herbal, where "Elleborus albus" = "Pelletour of Spayne / or longe worte" (? i.e. lingwort). A GH equivalent is *whyte elebore:* "whyte elebore or peleter" (sub "De Afara"). For other references, see Britten & Holland, p. 373.

peny wort	Usually in English applied to *Umbilicus rupestris*, a plant not implied in "Ameos", however (see Camus 1886, p. 34, and Fischer 1929, p. 259, and cf. Turner 1548 sub "Ami"). A GH equivalent of *peny wort* is *woodnep*. *Umbilicus rupestris* is described in GH sub "De Cotilidion": "Cotilidion is an herbe ... and is called timbalaria / and vmbelicus veneris. It hath rounde leues and thycke / and groweth on coueringes of olde buyldynges."
percely	Variant GH form: *parcyly*.
perwynke	See p. 34.
pimpinell	Cf. Camus 1886, p. 104. For GH forms, see p. 43.
pireter	Equivalent of *walworde*, the pellitory of the wall. Cf. OED pyrethrum and Grigson 1974, p. 166. Cf. also above sub *pellyter*.
policary	See Brodin 1950, p. 286.
pomgarnades	Variant GH forms: *(pome)garnat(e)*, *pomgarnatis*, *pomegarnette(s)*.
pompon	"Melons that we call pompons."
porcelayne	For GH forms, see p. 43.
prestes hode	See p. 38.
pyllulary	Cf. sub *lyse grasse*.
pyne apples	Cf. Britten & Holland, p. 381.
pyony	Variant GH form: *peonie*.
rasyns of carans	i.e. raisins from Corinth. Cf. Grigson 1974, p. 65.
rede	ME as plant name.
reed brere	ME as *Rosa canina*.
reed madder	As in Banckes's herbal. Cf. also Brodin 1950, pp. 253 and 289.
remcope	See p. 41.
rewbarbe	Variant GH form: *ruberbe*. "there be two maners thereof. One is called Reubarbarū ... The other rewponticum / bycause it groweth in an yle called ponticum ... and that is called rewpontyke".
ruddes	Cf. Britten & Holland, p.409.
rue	Variant GH form: *rew*.
rue of the felde	OE *feldrūde*. Equivalent of *wylde rue* ("Rue ... is in two maners. That is tame and wyld / yᵉ wylde is called pyganium").
ryb(b)wort	OE *ribbe*.
saffron	"Crocus is saffron / and there be two kyndes / one is named crocus ortensis / that is gardyn saffron .... The other is called orientalis saffron of orient / bycause it groweth in the eest ... This orient saffron is put in vomite medycyns."
sapyn (tre)	"Terbentyne ... is the gomme of a tree called sapyn or fyrre." Cf. Camus 1886, p. 125, and OED sapin ("a kind of fir or pine").
sarazyns mynt	"There is yet an other mynte / and it is called mynte romayne / or sarazyns mynt." A GH equivalent is *wytmynt*, i.e. *white mint* ("De menta romana. Wytmynt"), as

is *sarazyne:* "Mynte romayne or sarazyne is hote and drye in the seconde degre." (Cf. OED.) Possibly, *Mentha gentilis* is intended in GH. See Fischer 1929, p. 275, and Marzell 3 (1977), 154. "Menta romana" is however also an old name for *Tanacetum balsamita (Balsamita major).* See Camus 1886, p. 91, Fischer 1929, p. 286, and Marzell 4 (1979), 574.

sauyn	"Sauyne . . . is comynly had in religious cloysters."
saxifrage	The picture in GH (as in *Le grant herbier*) shows a fern. Britten & Holland (p. 416) identifies the plant as *Ceterach officinarum.* Possibly *Asplenium.* Cf. Marzell 1 (1943), 487, 490.
saynt peterworte	Cf. Britten & Holland, p. 412.
scabyous	In Turner 1965 identified as *Centaurea scabiosa.* Cf. Britten & Holland, p. 417, Frisk 1949, p. 221, and OED.
scaryole	"Endiuia . . . is other wyse called scaryole" (De endiuia); "scaryole that is wylde letuse" (De Boragine).
see onyon	Also called (sub "De squilla") "Onyon or chyboll of the see". (< Fr. "oignon ou cibole marine".) Cf. also *squyll* below.
selfegrene	For the formal variation of this name, see OED and Grigson 1975, p. 198. OE *singrēne.*
senacion	"Senacions is cresses / whan receptes expresseth senacions in the plurell nombre / it is to wyte cresses. But yf senacion be wryten in the synguler nombre / it is an other herbe that shall be spoken of after warde" (i.e. *Senecio vulgaris*).
senechon	See OED sencion.
seneuy	Cf. OED senvy and Frisk 1949, p. 301.
s[p]ereworde	See p. 43. The picture shows *Ranunculus sceleratus* rather than *R. flammula.*
setwall	Variant GH form: *setwale* (for *Curcuma*). The assignment of *setwall* to *Trigonella* ("Fenegreke or setwall") seems to be peculiar to GH. Usually, the name is applied to *Valeriana pyrenaica* (first in Turner 1538).
skyrwyt	Sub "De bursa pastoris", *skyrwit the lesse* (< Fr. *eruque petite*) is distinguished (probably the same plant as implied in "Eruca siluestris" sub "De Eruca"; cf. *wylde cawles* below). For the bot. ref. of GH *skyrwyt* cf.: "It is also called pastinaca" (De baucia) and "Skyrwyt. Or wylde cawles that bered mustarde sede" (De Eruca). Cf. Camus 1886, p. 40, Fischer 1929, p. 277, and Frisk 1949, p. 217.
sloe	"Achasia . . . is the iuce of sloes vnrype and wylde."
smerewort	Last OED quotation from "c. 1450" (*Alphita*).
snakegrass	"Dragons or snakegrass". Cf. *serpentyne.* VBD: *snakes grasse.*
sorbes	"Sorbes is the fruyte of a tree that is good to eate."
sorell de boys	"sorell de boys or cukowes mete".
sowthistle	As in VBD. *Sowthistle* appears in the heading, but not in the text, where *holy thystle* is used. Usually, *sowthistle* denotes *Sonchus oleraceus* (see e.g. Britten & Holland, p. 444).

sperage	Variant GH forms: *sparge, sperache.*
spourge	Also spelt *spurge* in GH.
spyknarde	Variant GH forms: *spyknade, spynarde.* "There be two maners of spyke / one is spyknarde / and ye other spyke celtyk."
stammarche	For *strammarche* in GH, see p. 43.
strawberye(s)	The plural form is used about the fruit, the singular about the plant ("Fragaria is an herbe called stra[w]bery").
strofulary	Identity obscure.
styche wort	Cf. *goosbyll.* In Banckes's herbal "Stychworte or Byrdestonge".
swete commyn	"[Anisum] is also called swete commyn / and it is the sede of an herbe so called."
swynefenell	Equivalent of *dogfenell* ("dogfenell or swynefenell").
swynes brede	"a rote called malum terre or swynes brede" (De Apio). See Camus 1886, p. 54 ("Ciclamen"), and Fischer 1929, pp. 260 and 266. Cf. *hogges meate.* Another English name (not in GH) is *sow-bread* (OED first record from c. 1550). In Franckenius 1659 *Swinbrödh* (i.e. swine bread).
swynes grasse	"poligonia y^t is swines grasse" (De camomylla) / "knotwort or swynes grasse" (Coronaria). Cf. Britten & Holland, p. 229, and Bierbaumer 1979, p. 225.
synkefoyle	A name of great formal variability. Banckes's herbal: *quynckfoyle,* Turner 1538: *synckfoly,* Turner 1548: *cynkfoly,* Gerard 1597: *cinkefoile.* In OE (the Latin) *quinquefolia* (Bierbaumer 1975, p. 115).
tamaryte	Variant GH forms: *tamaryke, tamaryst.*
tame cresses	Many GH 'plant names' (usually, with vague plant reference, reflexes of designations in the French text) consisting of *tame* + noun seem to be first recorded in GH, as *tame cress, tame onion, tame rue, tame lily, tame mint.* Cf. OED tame a. 2 (records only from 1551 onwards). Cf. list at the end of the Overall list.
tapsebarbe	See OED.
tasyl	Usually applied to *Dipsacus.* Cf. *wylde tasyll* below.
tintymall	Variant GH forms: *tintinall, titymall (tytimall).*
tormentyll	Cf. Camus 1886, p. 33, and Fischer 1929, p. 106. The association of *tormentil* with "De Albatra" (*Arbutus*) is mysterious.
tutson	In the text, *Vitex agnus-castus,* a mediterranean / central Asian shrub, is described. The use of *tutson* here is an example of a familiar plant / name / being "substituted for an unknown mediterranean herb under the old classical name" (Morton 1981, p. 98). Cf. also Grigson 1974, p. 222.
valerian	Cf. Camus 1886, p. 68 (*Valeriana officinalis*). For the identification of *valerian* as *Polemonium cæruleum,* see Britten & Holland, p. 233 ("Greek Valerian") and Turner 1548 (1965). For OE records, see Bierbaumer 1975, p. 122.
walfarne	Usually taken to be *Polypodium vulgare* (OED / Britten &

	Holland, p. 180), but the concurrent synonymous GH names indicate some species of *Asplenium* (cf. Camus 1886, p. 106).
walworde	Variant GH form: *walwort(e)*. Cf. Banckes's herbal: "Ebulus. This is called walworte."
wartwort	Britten & Holland (p. 484) identifies GH *wartwort* as *Coronopus squamatus* (wartcress), for which *nasturtium verrucarium* is an old name (Marzell 1 (1943), 1186). But GH *herucaria* may stand for *verrucaria* = *Euphorbia*, for which *wartwort* is an old English name. Cf. OED wartwort, Marzell 2 (1972), 363, Mowat 1887, and Turner 1548 (1965).
water flagge	Cf. *gladon*.
whyte elebore	Cf. *elebore* and *peleter of Spayne*.
whyte vyne	"bryony or whyte vyne".
wolfe thystle	Cf. Banckes's herbal: "woluysshe thystel or a wylde thystell".
woodnep	"Woodnep / or penywort". Cf. *penywort*.
woodroue	For identification, see Camus 1886, p. 39, Fischer 1929, p. 106, Bierbaumer 1975, pp. 146 f., 1979, pp. 267 f., and OED woodruff.
woodsorell	*Woodsour* is ME. Cf. p. 49.
woodyp	See p. 41.
wylde aloes	"a wood or tre named Camelia".
wylde bourache	See Brodin 1950, p. 302 (not in OED), and cf. *buglosse*.
wylde cawles	In Turner 1548 (as in MED) for *Sinapis arvensis* (cf. *skyrwyt*). Cf. Franckenius 1659: "Eruca sylv. Wild senap" (i.e. wild mustard).
wylde cowcomers	For identification, see Turner 1965 and MED.
wylde cresses	"Groweth about hye wayes". Turner (*Herball*) parallels *wild cress* with *waycress* and *sciatic cress* (possibly some species of *Lepidium*; see Britten & Holland, p. 129). In Lyte 1578 and Gerard 1597 *wild cress* seems to imply *Thlaspi alpestre*. Cf. Franckenius 1659: "Thlaspi. *Nasturtium agreste* ... Wäghkrassa" (i.e. waycress).
wylde galyngale	Only in the "Registre", sub "Ciperus". Most probably *Cyperus longus* is intended (cf. Wessén 1924, p. 68).
wylde garlyke	For identifcation, see Fischer 1929, p. 258, Turner 1965 and MED.
wylde hempe	See Brodin 1950, pp. 257 and 302.
wylde letuse	In MED identified as "? Lactuca virosa or L. scariola" (= *L. serriola*).
wylde malowes	In ME (MED) *wild mallow* denotes "either the common mallow or the marsh mallow growing wild" (i.e. *Malva sylvestris* or *Althæa officinalis*). Cf. Brodin 1950 (= *Althæa*) and Turner 1548 (= *Malva sylvestris*).
wylde mynte	Equivalent of *horsmynte* (which see).
wylde neppe	"Wylde neppe or bryony" (as in Turner 1538/1548).
wylde percely	See Brodin 1950, pp. 285 and 303, and Turner 1548 (1965).

96

	Camus 1886 (p. 118) and Fischer 1929 (p. 108) suggest *Aethusa cynapium*.
wylde rapes	See OED wild rape (Turner 1551) and cf. Wessén 1924, p. 80.
wylde rue	Or *rue of the felde* (see above). Cf. Fischer 1929, p. 282, and Marzell 2 (1972), 1552.
wylde sawge	Cf. *eupatory* and *hyndhele*.
wylde smalache	"Groweth in water". For possible bot. ref., see Camus 1886, p. 37, and Fischer 1929, p. 106 ("Apium raninum"). The 1561 ed. of GH gives as English names "wylde smalache or gowlandes".
wylde tasyll	Cf. Turner 1548 (1965).
wylde tyme	In GH, *Thymus serpyllum* is probably intended. But the old sense of *polium montanum* is *Teucrium polium*. Cf. Fischer 1929, p. 286, Marzell 4 (1979), 672, and Wölfel 1939, p. 140.
wylde valeryan	Also "called feu or valerian". The plant referred to sub "De silfu" in *Circa instans* and *Le grant herbier* seems to be *Thalictrum faetidum* (as also reflected in the English text: "and it stynketh"). See Camus 1886, p. 117, and Fischer 1929, p. 108.
wylde wyne	Cf. sub "De Viticella", where *wylde vyne* (ME) may denote *Bryonia*: "Viticella is a wede that is lyke a wylde vyne or gourde" (< Fr. "qui est sēblable a brionne"). Cf. Britten & Holland, p. 480.
wytmynt	See *sarazyns mynt*. In the "Registre" *whyte mynte*.
yelowe flagge	The Latin heading suggests *Iris foetidissima*—cf. Marzell 2 (1972), 1020, and Gerard 1597 (p. 53): "Stinking Gladdon is called in Latine *Spatula faetida*"—but this plant has normally (grey-) purple flowers. The name in VBD for *Iris pseudacorus* is *yelow lylles*.

Appendix to the Overall list

GH plants / fruits as characterized in terms of "common", "garden" / "tame" or "wild" (lists not exhaustive):[1]

common	bean, calamint, fennel, garlick, hemp, onion, smallage
garden / tame	brotherwort, cress, fig, garlick, ginger, lily, mallow, mint, mulberry, onion, parsley, pear, rue, saffron, sage, skirret
wild	aloe, apple, borage, brotherwort, cole, cucumber, fig, galingale, garlick, gourd, hemp, leek, letuse, lily, mallow, mint,

---

[1] Names like *wild nep*, where the simplex *nep* does not occur in GH, are excluded here. The comparatively great number of such specifications and of others (as exemplified below) in GH is chiefly due to the French original. In e.g. the contemporary Banckes's herbal, which is not a direct translation, such modifications are, on the whole, far less frequent.

mulberry, parsley, pear, poppy, rape, rue, sage, smallage, teasel, valerian

GH plants / fruits with some other specifications (denoting colour, form, height (size), taste):

> basil (gentle), almond (bitter, sweet), briar (red), bryony (black, white), bur (great, less, little), centaury (great), clote (little), consould (less, middle, more), cumin (sweet), fig (black, white), hellebore (black, white), hyssop (great, less), ivy (black), madder (less, more, red), mallow (high), mandrake (male, female),[2] melon (long, round), mint (white), mugwort (great, less, middle, small), nenuphar (white, yellow), onion (long, round, red, white), pepper (black, white, long), plantain (great, little, long), plum (black, red), policary (less, mean, more), poppy (black, red, white), saxifrage (less), skirret (less), woodbine (less, mean, more, yellow; sub "De Uolubilis").

[2] As e.g. with agaric, hare's beard and hig(h) taper (*Verbascum*).

Modern Scientific Names and the Corresponding English Names in The Grete Herball (1526)

Owing to the numerous uncertain plant references in GH the list is only tentative. As a rule, GH names are given in the singular. For GH variant forms (spellings), see the Overall list. For identification and synonymy, cf. pp. 45 ff. and Notes to the Overall list.

Acanthus mollis	bearefote, beares bough, beares twygge
Achillea millefolium	blodworte, carpenters grasse, myllefoyle, sanguinary, yarowe
Adiantum capillus-veneris	maydin here
Agrimonia eupatoria	egrymony
Agropyron repens	quekes
Agrostemma githago	cokyll, darnell, drawke, ray
Alchemilla	lyons fote, pedelyon
Alkanna tinctoria	alcamet
Allium cepa	onion
Allium porrum	leke
Allium sativum	churles tryacle, (tame) garlyke
Allium schoenoprasum	cyues
Allium ursinum	rampsons, wylde garlyke
Allium vineale	wylde garlyke (cf. also *wylde leke* in Overall list)
Aloe	aloe(n)
Alpinia officinarum	galyngale
Althaea officinalis	hye malowe, wylde malowe
Althaea rosea	garden (tame) malowe, hocke, holyhocke
Amaracus dictamnus	dytany, gardyn gynger
Anchusa (or "similar" plants)	buglosse, langdebefe, oxtongue, wylde bourache
Anethum graveolens	anet, dyll
Anthemis cotula	maydenwede, maythen, myddle consoulde
Anthriscus cerefolium	cheruell
Apium graveolens	smalache, stammarche
Arctium	clote, clyuer, grete burre, lesse burre, lytell burre
Aristolochia longa	reed mader
Aristolochia rotunda	meke galyngale, smerewort
Artemisia absinthium	wormwood
Artemisia vulgaris	the moder of herbes, moderwort, mugwort (cf. also *grete/lesse/small mugwort* in Overall list)

Arum maculatum	aaron, calues fote, cuckowe pyntyll, prestes hode
Asparagus officinalis	sperage
Asperula cynanchia	herbe (grasse) of vine, herbe squynantyke
Asphodelus	affodylly, woodroue
Asplenium	erththought, politryke, walfarne (cf. also *saxifrage* in Overall list)
Atriplex hortensis	arache
Atropa belladonna	dwale, more morell
Avena sativa	otes
Bellis perennis	bonewort, brusewort, daysy, lesse consoulde
Berberis vulgaris	berberies
Beta vulgaris	bete
Betonica officinalis	bethony
Borago officinalis	borage
Brassica oleracea	caule wort
Brassica rapa	rape
Brassica/Sinapis	lesse skyrwit, seneuy, wylde cawle, wylde rape
Bryonia alba	whyte bryony, whyte vyne
Bryonia dioica	bryony, wylde gourde, wylde neppe (cf. also note to *wylde wyne* in Overall list)
Buxus sempervirens	box /tre/
Calamintha	calamynt, nespyte
Calamintha officinalis	calamynt of the mountayne
Calendula officinalis	mary gowles, ruddes
Cannabis sativa	hempe
Capparis spinosa	capparis
Capsella bursa-pastoris	cassewed, sanguinary, shepeherds purs
Carduus	thystle
Carlina	wolfe thystle
Carum carvi	carui
Cassia acutifolia	sene
Castanea sativa	chestayne, chestnutte
Centaurea calcitrapa	caltrappe
Centaurea scabiosa	scabyous (for *Centaurea* cf. also *grete century* in Overall list)
Centaurium erythraea	centory, erthe galle
Ceterach officinarum	ceterach, saxifrage (cf. Asplenium above)
Chamaemelum nobile	camomylle
Chelidonium majus	cellydony
Chenopodium bonus-henricus	mercury
Cicer arietinum	chyches
Cichorium endivia	endyue, scaryole
Cichorium intybus	chycory
Cinnamomum	canell, cynamome
Cirsium	thystle
Citrullus colocynthis	colloquintide (colloquintida), gowrde of Alexandry, wylde gourde
Citrus limon	lymon

Citrus medica	cytron
Citrus sinensis	orenge
Cnicus benedictus	holy thystle, sowthistle
Cocos nucifera	nutte of Inde
Commiphora opobalsamum	bawme /tre/
Conium maculatum	hemlocke
Convolvulus	woodbynde
Coriandrum sativum	coryandre
Coronopus squamatus	buckesshorne, wartwort
Corylus maxima	auelane, fylberde
Crithmum maritimum	croyt marine (waterwort in GH 1561)
Crocus sativus	gardyn saffron, orient saffron, saffron
Cucumis citrullus	cytrulle
Cucumis melo	melon, pompon
Cucumis sativus	coucommer, cowgourde
Cucurbita pepo	cytrulle, gourde
Cuminum cyminum	comyn
Cupressus sempervirens	cypresse
Curcuma zedoaria	setwall
Cuscuta	dodyr
Cyclamen europaeum	hogges meate, swynes brede
Cydonia oblonga	quynce /apple/
Cymbopogon schoenanthus	camelles strawe, squinant
Cynoglossum officinale	hondes tonge
Cyperus	rysshe
Cyperus longus	thre cornerde rysshe, wylde galyngale
Dactylorhiza (see Orchis)	
Daphne laureola	laureole
Daucus carota	dawke
Delphinium staphisagria	lyse grasse, pyllulary
Dianthus caryophyllus	gylofre
Digitaria sanguinalis	goosfote, sanguinary
Dipsacus fullonum ssp. fullo-num	wylde tasyll
Dracunculus vulgaris	dragons, serpentine /of dragons/, snakegrass
Dryopteris filix-mas	heferne
Ecballium elaterium	wylde cowcomer
Equisetum	mares tayle
Eruca sativa	skyrwyt
Eryngium maritimum	reed brere, tasyll, thystle of the see, yringe
Eugenia aromatica	clowes
Eupatorium cannabinum	wylde hempe
Euphorbia	spourge, tintymall /of babylon/ (cf. also *wartwort* in Overall list)
Euphrasia	eufrace
Ficus	fygge
Filipendula vulgaris	dropwort
Foeniculum vulgare	fenell
Fragaria vesca	strawberye
Fraxinus excelsior	asshe tre

Fumaria officinalis	fume (smoke) of the erthe, fumyterry
Galium aparine	clyuers
Galium odoratum	woodroue
Gentiana	gencyan
Gentianella	baldymony, felwort
Geranium	doues fote
Geum urbanum	auens, gylofre
Glycyrrhiza glabra	lycoryce tre
Gossypium	cotton
Hedera helix	blacke yuy, yuy
Helleborus niger	(blacke) elebore, lyons fote, pedelyon
Helychrysum stoechas	Hercules grasse
Hieracium pilosella	mows eare
Hordeum vulgare	barley
Humulus lupulus	hoppes
Hyoscyamus niger	henbane
Hypericum perforatum	herbe John, saynt Johannis (Johns) wort
Hyssopus officinalis	ysope
Inula helenium	elfe docke, horshele, scabwort
Iris germanica	blewe flourdelyce
Iris pseudacorus	gladon, water flagge, yelowe flagge (cf. also *flag* in Overall list)
Juglans regia	wall nutte
Juncus	rysshe
Juniperus communis	jenepre
Juniperus sabina	sauyn
Lactuca sativa	letuse
Lactuca serriola	scaryole, wylde letuse
Lamium	archaungell, blynde nettell, deed nettel
Lavandula angustifolia	lauendre
Laurus nobilis	bayes, laurel
Lawsonia inermis	alcamet
Lemna minor	ducke meate, frogges fote, grenes, lentylles of the water
Lens culinaris	lentyle
Lepidium sativum	cress, gardyne (tame) cresse
Levisticum officinale	louage
Lilium candidum	lylly
Linum usitatissimum	flax, lyne
Lithospermum officinale	gromyll, lychwale, lychworte
Lobaria pulmonaria	crayfery, lungwort
Lolium temulentum	cokyll, ray, (darnel)
Lonicera periclymenum	cheruell, gotes leues, woodbynde
Majorana hortensis	Margetym gentyll, mariorayne
Malus	apple
Malus sylvestris	wood crabbe, wylde apple, wyldynge
Malva	malowe
Mandragora officinarum	mandragora, mandrake
Marchantia polymorpha	lyuerwort
Marrubium vulgare	horehounde

Melilotus	honysocle, kynges crowne, mellilot
Melissa officinalis	bawme /tre/, melisse
Mentha gentilis	sarazyns mynt, wytmynt
Mentha longifolia	horsmynte, wylde mynte
Mentha spicata	gardyn (tame) mynte, mynte
Mespilus germanica	medler, mespile, open arse (for *nefle*, see p. 51)
Meum athamanticum	meu
Morus nigra	molberye
Musa paradisiaca	apple of paradys
Myristica fragrans	nutmygge
Myrtus communis	myrte (for *myrtylle*, see note in Overall list)
Nardostachys jatamansi	spyke, spykenarde
Nerium oleander	oleandre
Nigella sativa	cokyll, gith
Nuphar/Nymphaea	nenufar
Ocimum basilicum	basyll
Olea europaea	olyue
Orchis (Dactylorhiza)	gangelon, hare ballockes, satirion
Origanum majorana (see Majorana hortensis)	
Origanum vulgare	brotherworte, orygan
Orysa sativa	rys
Oxalis acetosella	alleluya, cuckowes brede, cukowes meate, sorell de boys, woodsorell
Paeonia	pyony
Panicum mileaceum	mylle, myllet
Papaver	poppy
Papaver rhoeas	reed poppy, wylde poppy
Papaver somniferum	blacke poppy, white poppy
Parietaria diffusa	pireter, walworde
Pastinaca sativa	skyrwyt
Petroselinum crispum	percely
Peucedanum officinale	dogfenell, swynefenell
Phoenix dactylifera	date
Phragmites communis	rede
Phyllitis scolopendrium	hertes tongue
Picea/Pinus	fyrre, pyne tree, sapyn /tre/
Pimpinella anisum	anys, swete commyn
Pimpinella saxifraga	pimpinell (pympernell), selfe heale
Piper nigrum	peper
Plantago lanceolata	longe plantayn, lytell plantayn, rybwort
Plantago major	grete plantayne, plantayne, weybrede
Platanthera (see Orchis)	
Polemonium caeruleum	valerian
Polygonatum multiflorum	our ladyes seale, Salamons seale
Polygonum aviculare	knotgrasse, knotwort, sentynode, sparow tongue, swynes grasse
Polygonum bistorta	bistorte

Polygonum hydropiper	arssmert, blodworte, culrage, sanguinary
Polypodium vulgare	oke ferne, polipodi
Polyporus officinalis	agaryk
Portulaca oleracea	porcelayne
Potentilla erecta	tormentyll (for another GH reference of *tormentyll*, see Overall list)
Potentilla reptans	fyue (.v.) leued /grasse/, synkefoyle
Poterium sanguisorba	lesse saxifrage
Primula veris	artetyke, cowslyp, herbe paralysy, pagle
Primula vulgaris	paralysy, prymerolle, saynt peterworte
Prunus (kinds of cherry)	agryote, cherye, damacene
Prunus Amygdalus	almonde (bytter, swete)
Prunus domestica	damasson, damaske plomme, plomme
Prunus domestica ssp. insititia	bolays
Prunus persica	peche
Prunus spinosa	sloe
Pteridium aquilinum	ferne
Pulicaria	policary
Punica granatum	pomgarnade
Pyrus	(tame, wylde) pere
Quercus robur	oke(n) tree
Ranunculus ficaria	bryght, celendyne
Ranunculus flammula	s[p]ereworde (cf. note in Overall list)
Ranunculus sceleratus	ache, crowfote
Raphanus sativus	radysshe
Rheum	rewbarbe
Rhinanthus	mederacle
Rorippa	water cress
Rosa	brere, rose, thorn
Rosa rubiginosa	eglentyne
Rosmarinus officinalis	rosmary
Rubia tinctorum	madder, warence
Rubus fruticosus	bramble, /wylde/ blacke beryes, wylde molberyes
Rumex	docke, reed docke
Rumex acetosa	sorell
Ruscus aculeatus	kneholme
Ruta graveolens	rue
Ruta montana	rue of the felde, wylde rue
Saccharum officinarum	sugre rede
Salix	wyloue /tree/
Salvia officinalis	sawge
Salvia sclarea	clarey
Sambucus ebulus	walworde
Sambucus nigra	eldre
Sanguisorba (see Poterium)	
Saponaria officinalis	burit, crowsoppe, fullers grasse, herbe phylyp, saponary
Sarothamnus scoparius	brome

Satureja hortensis	sauerey
Secale cereale	rye
Sempervivum tectorum	howsleke, jōbarde, selfegrene
Senecio vulgaris	grownswell, senacion, senechon
Setaria italica	panyke
Sinapis (see Brassica)	
Sison amomum	wylde percely
Smyrnium olusatrum	alysamder, percely of Macedony, stammarche
Solanum nigrum	lesse morell, nyght shade, petymorell
Sorbus domestica	sorbe
Spinacia oleracea	spynache
Stellaria holostea	goosbyll, styche wort
Strychnos nux vomica	spewynge nutte
Succisa pratensis	deuylles bytte, remcope
Symphytum officinale	comfrey, more consoulde
Tamarindus indica	date of Ynde
Tamarix gallica	tamaryte
Tamus communis	blacke bryony
Tanacetum balsamita	cost, costmary
Tanacetum vulgare	tansey
Taxus baccata	ewe
Teucrium chamaedrys	germaundre
Teucrium scorodonia	eupatory, hyndhele, wylde sawge
Thymus serpyllum	pellyter, wylde tyme
Trifolium	hare trefle, thre (.iii.) leued grasse, trefle
Trigonella foenum-graecum	fenegreke, setwall
Triticum aestivum	wheate
Tussilago farfara	lytell clote
Urginea maritima	chybol of the see, see onyon, squyll, water onion
Urtica	nettle
Valeriana	valerian
Valeriana celtica	spyke celtyk
Veratrum album	lyngwort, peleter of Spayne, (whyte) elebore
Verbascum thapsus	hare berde, hygtaper, moleyne, tapsebarbe
Verbena officinalis	veruayne
Vicia faba	beane
Vinca minor	perwynke
Viola odorata	vyolette
Vitex agnus-castus	tutson
Vitis labrusca	wylde wyne
Zingiber officinale	gynger

Bibliography

Anderson, F. J. 1977. *An Illustrated History of the Herbals.* New York.

Arber, Agnes. 1912. *Herbals, their Origin and Evolution.* Cambridge (2nd ed. 1938).

Arber, Agnes. 1941. "On Grasses in Herbal Literature". *Darwiniana* 5 (pp. 20–30).

Arber, Agnes. 1953. "From Medieval Herbalism to the Birth of Modern Botany." In: *Science, Medicine and History. Essays ... Written in Honour of Charles Singer.* I (pp. 317–36). London.

Arbolayre contenāt la qualitey et vertus. proprietey des herbes. arbes. gōmes. et semēces extrait de pluseurs tratiers de medicine [c. 1486. Besançon.]

Bailey, R. W. 1978. *Early Modern English. Additions and Antedatings to the Record of English Vocabulary 1475–1700.* Hildesheim & New York.

Banckes's Herbal. 1525. London.

Banckes's Herbal. 1941. Ed. S. V. Larkey & Th. Pyles. New York.

Barber, C. 1976. *Early Modern English.* London.

Bartholomæus Anglicus. 1975. *On the Properties of Things.* John Trevisa's translation of Bartholomæus Anglicus *De Proprietatibus Rerum.* A Critical text I–II (ed. M. C. Seymour). Oxford.

Bauhin, C. 1623. *Pinax theatri botanici.* Basileae.

Beck, C. H. 1940. *Studien über Gestalt und Ursprung des Circa Instans.* Würzburg.

Bentham, G. [& Hooker, J. D.]. 1858 [1887]. *Handbook of the British Flora.* London.

Bentley, R. & Trimer, H. 1880. *Medicinal Plants.* I–IV. London.

Bergh, B. 1978. *Paleography and Textual Criticism.* Regiae Societatis Humaniorum Litterarum. Scripta Minora Lundensis. Lund.

Bierbaumer, P. 1975–79. *Der botanische Wortschatz des Altenglischen.* 1–3. Frankfurt am Main.

Blunt, W. 1950. *The Art of Botanical Illustration.* London.

Blunt, W. & Raphael, S. 1979. *The Illustrated Herbal.* London.

Bock, H. 1539. *Kreuter Buch.* Strassburg.

Britten, J. & Holland, R. 1878–86. *A Dictionary of English Plant-Names.* London.

Brodin, G. 1950. *Agnus Castus. A Middle English Herbal.* Uppsala.

Brunfels, O. 1530–36. *Herbarum vivae eicones.* Strassburg.

Camus, G. 1886. *L'opera salernitana 'Circa Instans' ed il testo primitivo del 'Grant Herbier en Francoys'.* Modena.

Clapham, A. R., Tutin, T. G. & Warburg, E. F. 1962 (2nd ed.). *Flora of the British Isles.* Cambridge.

Cockayne, O. 1961. *Leechdoms, Wortcunning and Starcraft of Early England.* Revised ed. with a new introduction by Charles Singer. London.

Debus, A. G. 1978. *Man and Nature in the Renaissance.* Cambridge.

Dictionary of National Biography. 1885–. Ed. L. Stephen *et al.* London.

Dobson, E. J. 1968 (2nd ed.). *English Pronunciation 1500–1700.* Oxford.

Dony, J. G. *et al.* 1974. *English Names of Wild Flowers: a list recommended by the Botanical Society of the British Isles.* London.

Earle, J. 1880. *English Plant Names from the Tenth to the Fifteenth Century.* Oxford.

Fischer, H. 1929. *Mittelalterliche Pflanzenkunde.* München.

Fisher, R. 1932–34. *The English Names of our Commonest Wild Flowers.* Arbroath.

Franckenius, J. 1638 and 1659. *Speculum botanicum /renovatum/.* Upsaliae.

Freeman, Margaret, B. 1943. *Herbs for the Medieval Household.* New York.

Fries, S. 1962. "Flora svecica som växtnamnsordbok". *Svenska Linnésällskapets årsskrift* 45 (pp. 34–45).

Fries, S. 1975. *Svenska växtnamn i riksspråk och dialekt.* Umeå.

Fries, S. 1977. *Linné och de svenska växtnamnen. Ett kapitel ur det svenska växtnamnsskickets historia.* Umeå.

Fries, S. 1980. "Uppgifter inom svensk växtnamnsforskning". *Saga och sed* (pp. 28–39).

Fries, Th. M. 1904. "Svenska växtnamn. 1. Under medeltiden". *Arkiv för botanik.* Band 3. N:o 14 (pp. 1–60).

Frisk, G. 1949. *A Middle English Translation of Macer Floridus de Viribus Herbarum.* Uppsala.

Fuchs, L. 1542 & 1545. *De historia stirpium.* Basileae.

Gerard, J. 1597. *The Herball or Generall Historie of Plantes.* London (2nd ed. 1633 ed. by Th. Johnson).

Gerth van Wijk, H. L. 1911–16. *A Dictionary of Plantnames.* The Hague.

Le grant herbier. c. 1500. Paris (eds. of 1513(?), 1540(?) and 1545 also seen).

Green, J. R. 1914. *A History of Botany in the United Kingdom from the Earliest Times to the End of the 19th Century.* London.

Green, Mary L. 1927. "History of Plant Nomenclature". *Kew Bulletin* (pp. 403–15).

The Grete Herball. 1526, 1529, 1539, 1561. London.

Grigson, G. 1955 (1975). *The Englishman's Flora.* London.

Grigson, G. 1974. *A Dictionary of English Plant Names.* London.

Hegi, G. *et al.* 1906–. *Illustrierte Flora von Mittel-Europa.* München.

Heller, J. L. 1964. "The Early History of Binomial Nomenclature". *Huntia* 1 (pp. 33–70).

Henrey, Blanche. 1975. *British Botanical and Horticultural Literature before 1800.* I–III. London.

Hesselman, B. 1935. *Från Marathon till Långheden. Studier över växtnamn och naturnamn.* Stockholm.

Høeg, O. A. 1976. (3rd ed.). *Planter og tradisjon.* Oslo.

Hoeniger, F. D. & Hoeniger, J. F. M. 1969 (a). *The Development of Natural History in Tudor England.* Published for The Folger Shakespeare Library by The University Press of Virginia.

Hoeniger, F. D. & Hoeniger, J. F. M. 1969 (b). *The Growth of Natural History in Stuart England from Gerard to the Royal Society.* Published for The Folger Shakespeare Library by The University Press of Virginia.

Hudson, W. 1762. *Flora anglica.* Londini (2nd ed. 1778).

Huguet, E. 1925–67. *Dictionnaire de la langue française du seizième siècle.* 1–7. Paris.

Hulton, P. & Smith, L. 1979. *Flowers in Art from East and West.* London.

Kärre, K. 1924. "The English Plant-Name *groundsel*". *Studier i modern språkvetenskap* 9 (pp. 69–78).

Klein, E. 1966–67. *A Comprehensive Etymological Dictionary of the English Language.* I–II. Amsterdam.

Lange, J. 1959–61. *Ordbog over Danmarks Plantenavne.* 1–3. København.

Lange, J. 1966. *Primitive plantenavne og deres gruppering efter motiver.* København.

Larkey, S. V. & Pyles, Th. (see Banckes's Herbal).

Lawrence, G. H. M. 1965. "Herbals, their History and Significance". In: *History of*

Botany. Two papers presented at a Symposium held at the William Andrews Clark Memorial Library, December 7, 1963. Los Angeles.

Le Strange, R. 1977. *A History of Herbal Plants*. London.

Linnaeus, C. 1753. *Species plantarum*. Holmiae.

Longeon, C. 1980. "L'usage du latin et des langues vernaculaires dans les ouvrages de botanique du xvième siècle." In: *Acta conventus neo-latini turonensis* II (pp. 751–66). Ed. J.-C. Margolin. Paris.

Lyons, J. 1968. *Introduction to Theoretical Linguistics*. Cambridge.

Lyons, J. 1977. *Semantics*. 1–2. London.

Lyte, H. 1578. *A Nievve Herball, or Historie of Plantes*. Antwerpe.

Lyttkens, A. 1904–15. *Svenska växtnamn*. 1–3. Stockholm.

McClintock, D. & Fitter, R. S. R. 1955 (1971). *The Pocket Guide to Wild Flowers*. London.

Marzell, H. 1943–79. *Wörterbuch der deutschen Pflanzennamen*. Leipzig.

Middle English Dictionary. 1952–. Ed. H. Kurath, S. M. Kuhn and R. E. Lewis. Ann Arbor.

Morton, A. G. 1981. *History of Botanical Science*. London.

Mowat, J. L. G. (ed.) 1887. *Alphita. A Medico-Botanical Glossary*. Oxford.

Nissen, C. 1966 (2. Aufl.). *Die botanische Buchillustration*. Stuttgart.

Nordhagen, R. 1954. *Kusymre, kodriver, cowslip og paigle. Studier over gamle Primula-navn i Nordvest-Europa*. Oslo.

Nordhagen, R. 1965. *Botaniske studier over bringebær og andre folkelige Rubus-navn i Norden*. Oslo.

Onions, C. T. A. 1966. *The Oxford Dictionary of English Etymology*. Oxford.

The Oxford English Dictionary (with Supplements). 1933–82. Oxford.

Parkinson, J. 1629. *Paradisi in sole paradisus terrestris*. London.

Parkinson, J. 1640. *Theatrum botanicum: The Theater of Plants*. London.

Prior, R. C. A. 1863. *On the Popular Names of British Plants*. London (2nd ed. 1870, 3rd ed. 1897).

Putnam, Clare. 1972. *Flowers and Trees of Tudor England*. London [from a MS dated 1520–30].

Raven, C. E. 1947. *English Naturalists from Neckam to Ray*. Cambridge.

Raven, C. E. 1950 (2nd ed.). *John Ray, Naturalist. His Life and Works*. Cambridge.

Ray, J. 1660. *Catalogus plantarum circa Cantabrigiam nascentium*. Cantabrigiae.

Ray, J. 1670. *Catalogus plantarum Angliae et insularum adjacentium*. Londini (2nd ed. 1677).

Ray, J. 1690. *Synopsis methodica stirpium britannicarum*. Londini (2nd ed. 1696, 3rd ed. 1724).

Ray, J. 1724 (1973). *Synopsis methodica stirpium britannicarum*. Editio tertia. Facsimile with an Introduction by William T. Stearn. The Ray Society. London.

Rayner, J. F. 1927. *A Standard Catalogue of English Names of our Wild Flowers*. London.

Reeds, Karen, M. 1976. "Renaissance Humanism and Botany". *Annals of Science* 33 (pp. 519–42).

Rohde, Eleanour Sinclair. 1922. *The Old English Herbals*. London.

Rydén, M. 1978 (a). *Shakespearean Plant Names. Identifications and Interpretations*. Stockholm.

Rydén, M. 1978 (b). "The English Plant Names in Gerard's *Herball* (1597)". In: *Studies in English Philology, Linguistics and Literature Presented to Alarik Rynell 7 March 1978* (pp. 142–50). Stockholm.

Rydén, M. 1981. "English Plant-Name Research". In: *Papers from the First Nordic Conference for English Studies, Oslo, 17–19 September, 1980* (pp. 374–84). Oslo.

Rydén, M. forthcoming. "Antedatings and Additions for the OED from *The vertuose boke of Distyllacyon of the water of all maner of Herbes* (1527)". *Notes and Queries.*

Schäfer, J. 1980. *Documentation in the O.E.D.* Oxford.

Singer, C. (see Cockayne).

Smith, J. E., Sir. 1824–36. *The English Flora.* I–V. London.

Stannard, J. 1974. "Medieval Herbals and their Development". *Clio Medica* 9 (pp. 23–33).

Stearn, W. T. (see Ray 1724 (1973) and Turner 1965).

Stracke, J. R. (ed.). 1974. *The Laud Herbal Glossary.* Amsterdam.

Tabernaemontanus, J. 1590. *Eicones plantarum seu stirpium.* Francofurti ad Moenum.

Turner, W. 1538. *Libellus de re herbaria novus.* Londini.

Turner, W. 1548. *The names of herbes in Greke, Latin, Englishe Duche & Frenche wyth the commune names that Herbaries and Apotecaries vse.* London.

Turner, W. 1551–68. *A New Herball.* London & Collen.

Turner, W. 1877. *Libellus de re herbaria novus.* Ed. B. D. Jackson. London.

Turner, W. 1881. *The Names of Herbes.* Ed. James Britten. London.

Turner, W. 1965. *Libellus* and *The Names of Herbes.* Ed. William T. Stearn *et al.* London.

Ullman, S. 1957 (2nd ed.). *The Principles of Semantics.* Oxford.

Ullman, S. 1962. *Semantics. An Introduction to the Science of Meaning.* Oxford.

The vertuose boke of Distyllacyon. 1527. London.

Vide, S.-B. 1962. "De provinsiella växtnamnen i Linnés skånska resa." *Svenska Linnésällskapets årsskrift* 45 (pp. 46–54).

Vide, S.-B. 1967. "De sydsvenska växtnamnen. Namngivningsmotiv och formkategorier." *Sydsvenska Ortnamnssällskapets årsskrift* (pp. 16–50).

Wessén, E. 1924. "Svenska växtnamn från 1500-talet." *Linköpings Bibliotheks handlingar.* Ny serie 4 (pp. 48–85).

Wilson, E. 1979. "An Unpublished Alliterative Poem on Plant-names from Lincoln College, Oxford, MS. Lat. 129 (E)." *Notes and Queries,* New Series, Vol. 26 (pp. 504–8).

Withering, W. 1776. *A Botanical Arrangement of all the Vegetables naturally growing in Great Britain.* London.

Wölfel, H. 1939. *Das Arzneidrogenbuch Circa Instans.* Berlin.

Wright, J. 1898–1905. *The English Dialect Dictionary.* 1–6. London.

Wyld, H. C. (1936) 1953 (3rd ed.). *A History of Modern Colloquial English.* Oxford.

Abbreviations

GH	The Grete Herball 1526
VBD	The vertuose boke of Distyllacyon 1527
B–H	Britten & Holland 1878–86
OE	Old English
ME	Middle English
PE	Present-day English
OF	Old French
Lat.	Latin
Med. Lat.	Medieval Latin
Sw.	Swedish
MS(S)	manuscript(s)
DNB	Dictionary of National Biography
MED	Middle English Dictionary
OED	The Oxford English Dictionary
ed.	edition, edited by
ex(x).	example(s)
sg.	singular
pl.	plural
sp.	species
ssp.	subspecies
v.	variety

DATE DUE			

DEMCO 38-297